D1122527

INTEGRATED PEST MANAGEMENT FOR

POTATOES

IN THE WESTERN UNITED STATES

WESTERN REGIONAL IPM PROJECT
UNIVERSITY OF CALIFORNIA STATEWIDE IPM PROJECT
UNIVERSITY OF IDAHO
WASHINGTON STATE UNIVERSITY
COLORADO STATE UNIVERSITY
OREGON STATE UNIVERSITY
MONTANA STATE UNIVERSITY

UNIVERSITY OF CALIFORNIA, DIVISION OF AGRICULTURE AND NATURAL RESOURCES
PUBLICATION 3316

WESTERN REGIONAL RESEARCH PUBLICATION 011

1986

Under the procedure of cooperative publication, this regional report
becomes, in effect, an identical publication of each of the participating
experiment stations and agencies and is mailed under the indicia of each.

PRECAUTIONS FOR USING PESTICIDES

Pesticides are poisonous and must be used with caution. READ THE LABEL BEFORE OPENING A PESTICIDE CONTAINER. Follow the label precautions and directions, including requirements for protective equipment. Use a pesticide only against pests specified on the label or in published University recommendations. Apply pesticides at the rates specified on the label or at lower rates if suggested in this publication. Pesticide laws and regulations change frequently; be sure the publication you are using is up to date.

Legal Responsibility. The user is legally responsible for any damage due to misuse of pesticides, including possible effects of drift, runoff, or residues.

Transportation. Do not ship or carry pesticides with food or feed in a way that allows contamination. Never transport pesticides in a closed passenger vehicle or cab.

Storage. Keep pesticides in original containers until used. Store them in a locked cabinet, building, or fenced area where they are not accessible to children, unauthorized persons, pets, or livestock. DO NOT store pesticides with food, feed, fertilizers, or other materials that may become contaminated by the pesticides.

Container Disposal. Dispose of empty containers carefully. Never reuse them. Keep containers away from children and animals. Never dispose of containers where they may contaminate water supplies or natural waterways. Consult local agricultural authorities for procedures for disposing of large quantities of empty containers.

Protection of Non-Pest Animals and Plants. Many pesticides are toxic to desirable animals, including honeybees, natural enemies, fish, domestic animals, and birds. Crops and other plants may also be damaged by misapplied pesticides. Take precautions to protect non-pest species from direct exposure to pesticides and from contamination due to drift, runoff, or residues. Certain rodenticides may pose a special hazard to animals that eat poisoned rodents.

Posting Treated Fields. For some materials, re-entry intervals are established to protect workers. Keep workers out of the field for the required time after application and, when required by regulations, post the treated areas with signs indicating the safe re-entry date.

Harvest Intervals. Some material or rates cannot be used in certain crops within a specified time before harvest. Follow pesticide label instructions and allow the required time between application and harvest.

Permit Requirements. Many pesticides require a permit from the county agricultural commissioner or other local authority before possession or use. When such materials are recommended in this publication, they are marked with an asterisk (*).

Processed Crops. Some processors will not accept a crop treated with certain chemicals. If your crop is going to a processor, be sure to check with the processor before applying a pesticide.

Crop Injury. Certain chemicals may cause injury to crops (phytotoxicity) under certain conditions. Always consult the label for limitations. Before applying any pesticide, take into account the stage of plant development, the soil type and condition, the temperature, moisture, and wind. Injury may also result from the use of incompatible materials.

Personal Safety. Follow label directions carefully. Avoid splashing, spilling, leaks, spray drift, and contamination of clothing. NEVER eat, smoke, drink, or chew while using pesticides. Provide for emergency medical care IN ADVANCE as required by regulation.

ORDERING

For information about ordering this publication, write to:

Publications
Division of Agriculture and Natural Resources
University of California
6701 San Pablo Avenue
Oakland, California 94608-1239

or telephone (415) 642-2431

Publication #3316

Other books in this series include:

Integrated Pest Management for Walnuts, Publication #3270
Integrated Pest Management for Tomatoes, Publication #3274
Integrated Pest Management for Rice, Publication #3280
Integrated Pest Management for Citrus, Publication #3303
Integrated Pest Management for Cotton, Publication #3305
Integrated Pest Management for Cole Crops and Lettuce, Publication #3307
Integrated Pest Management for Almonds, Publication #3308
Integrated Pest Management for Alfalfa Hay, Publication #3312

ISBN 0-931876-74-5
Library of Congress No. 86-70412

© 1986 by The Regents of the University of California
Division of Agriculture and Natural Resources

All rights reserved.
No part of this publication may be reproduced, stored in a retrieval system, or transmitted, in any form or by any means, electronic, mechanical, photocopying, recording, or otherwise, without the written permission of the publisher and the author.

Printed in the United States of America

6m-9/86

The University of California in compliance with the Civil Rights Act of 1964, Title IX of the Education Amendments of 1972, and the Rehabilitation Act of 1973 does not discriminate on the basis of race, creed, religion, color, national origin, sex, or mental or physical handicap in any of its programs or activities, or with respect to any of its employment policies, practices, or procedures. The University of California does not discriminate on the basis of age, ancestry, sexual orientation, marital status, citizenship, nor because individuals are disabled or Vietnam era veterans. Inquiries regarding this policy may be directed to the Affirmative Action Officer, Division of Agriculture and Natural Resources, 2120 University Ave., University of California, Berkeley, California 94720 (415) 644-4270.

Contents

Contributors and Acknowledgments

This manual was produced under the auspices of the University of California Statewide IPM Project with assistance and additional funding from the SAES Western Regional Research Project W-161, "Integrated Pest Management for Semi-arid, Dryland, and Irrigated Agroecosystems in the Western Region;" the University of Idaho; Washington, Colorado, Oregon, and Montana State Universities; and the California Potato Advisory Board.

Prepared by the IPM Manual Group of the Statewide IPM Project at the University of California, Davis, Mary Louise Flint, Director and Technical Editor.

Larry L. Strand, Principal Senior Writer
Paul A. Rude, Senior Writer (Insect Chapter)
Jack Kelly Clark, Principal Photographer

Technical Coordinators

Guy W. Bishop, Professor of Entomology, University of Idaho, Southwest Idaho Research and Extension Center, Parma

Gary D. Kleinschmidt, Extension Potato Specialist, University of Idaho Cooperative Extension Service, Twin Falls

Kenneth W. Knutson, Extension Potato Specialist, Colorado State University, Fort Collins

Alvin R. Mosley, Associate Professor and Potato Specialist, Oregon State University, Corvallis

Robert E. Thornton, Extension Horticulturist, Washington State University Cooperative Extension Service, Pullman

Ronald E. Voss, Extension Vegetable Specialist, University of California, Davis

Contributors

Abbreviations after contributors' names stand for these institutions: (ARS)—United States Department of Agriculture, Agricultural Research Service; (CAC)—California Agricultural Commissioner; (CS)—Colorado State University; (MS)—Montana State University; (MSA)—Montana State Department of Agriculture; (NM)—New Mexico State University; (NSA)—Nevada State Department of Agriculture; (ND)—North Dakota State University; (OS)—Oregon State University; (UA)—University of Arizona; (UC)—University of California; (UI)—Univeristy of Idaho; (US)—Utah State University; (UW)—University of Wyoming; (WS)—Washington State University.

Entomology:
Oscar Bacon (UC), Guy W. Bishop (UI), Chris Burkhardt (UW), Vernon E. Burton (UC), Whitney Cranshaw (CS), Don Davis (US), Lee Fox (ARS), Bob Gillespie (MSA), Hugh Homan (UI), Jeff Knight (NSA), Donnie Powell (ARS), Harold Toba (ARS)

Horticulture, Physiology:
John Guerard (UC), Larry K. Hiller (WS), W. M. Iritani (WS), John Kelley (OS), Gale E. Kleinkopf (UI), Gary D. Kleinschmidt (UI), Alvin R. Mosley (OS), J. J. Pavek (ARS), Robert E. Thornton (WS), Herman Timm (UC), Kent B. Tyler (UC), Ronald E. Voss (UC), Dale T. Westermann (ARS)

Nematology:
Saad L. Hafez (UI), Winfield Hart (UC), Harold J. Jensen (OS), Paul A. Koepsell (OS), Edward L. Nigh (UA), Jack Pinkerton (WS), Philip A. Roberts (UC), Gerald S. Santo (WS), Ivan J. Thomason (UC)

Plant Pathology:
Tom C. Allen (OS), R. Gary Beaver (UI), Tully Bowman (UC), Jim R. Davis (UI), Gene D. Easton (WS), Gary D. Franc (CS), John Guerard (UC), Dennis H. Hall (UC), Monty D. Harrison (CS), Gary D. Kleinschmidt (UI), Kenneth W. Knutson (CS), Paul A. Koepsell (OS), Clark Livingston (CS), Gary A. McIntyre (CS), Mary L. Powelson (OS), Gary A. Secor (ND), M. E. Stanghellini (UA), M. K. C. Sun (MS), Peter E. Thomas (ARS), A. R. Weinhold (UC)

Vertebrates:
Rex E. Marsh (UC), Jim Massey (CAC)

Weed Science:
Harry S. Agamalian (UC), Arnold Appleby (OS), Richard N. Arnold (NM), Roger Benton (UC), Paul Bessey (UA), Richard L. Chase (US), Lloyd C. Haderlie (UI), Eugene Heikes (CS), Harold M. Kempen (UC), Alex G. Ogg, Jr. (ARS), Robert Parker (WS), Charles C. Stanger (OS)

Special Assistance

In addition to the Technical Coordinators and many of the Contributors, the following people provided special assistance in obtaining photographs:
Walt Bentley, Dick Brewer, Edward Butler, Harry Carlson, Mary DeVoy, Richard Hom, Don Kirby, Mike Nagle, Joe Picard, Jerry Smalley, Albert Ulrich, Jim Whitmire, Bert Wilcox

Special Thanks

The IPM Manual Group extends special thanks to James M. Lyons, Director of the University of California Statewide IPM Project, Howard Ferris, Associate Director, and Gary A. McIntyre, Coordinator of the Western Regional Research Project W-161. Thanks also to Agriculture and Natural Resources Publications, University of California, for the assistance of the editor and artists.

The following persons have generously provided information, offered suggestions, or reviewed draft manuscripts:
D. O. Adams, L. Askham, D. Baker, E. Bechinski, W. J. Bentley, K. Bohnenblust, B. Brewster, B. Brodie, W. M. Brown, J. Bushnell, D. N. Byrne, R. H. Callihan, W. Callison, H. Carlson, R. Clarke, D. L. Corsini, D. Cudney, R. Davidson, S. A. Dewey, L. Disburgh, B. Dixon, R. B. Dwelle, M. English, D. Feuhring, G. Fisher, W. Foeppel, J. A. Fox, A. G. Fradkin, J. G. Garner, C. Glover, P. Goodell, A. S. Greathead, G. D. Griffin, D. C. Gross, O. Gutbrod, R. W. Hackney, J. L. Halderson, J. Hall, M. Jackson, D. James, D. Horton, M. Johnson, W. Johnson, A. Kelman, D. C. Larsen, R. D. Lee, S. Lindow, O. Lorenz, P. Marer, J. Maxfield, N. F. McCalley, S. L. Michener, E. Mulrean, S. D. Miller, J. H. O'Bannon, N. Oebker, D. C. Opgenorth, R. R. Romanko, D. Roth, L. Sandvol, M. Schroth, E. Shannon, H. Smith, W. C. Sparks, M. W. Stimmann, R. L. Stoltz, S. Thomas, S. Thomson, R. Todd, D. Viglierchio, A. C. Weiner, M. Workman, S. Young, M. Zavala

Information in this manual has been derived from research conducted and supported by Colorado, Montana, New Mexico, Oregon, Utah, and Washington State Universities, the Universities of Arizona, California, Idaho, Nevada, and Wyoming, the United States Department of Agriculture, SAES Western Regional Research Project W-161, and the potato industry.

Production

Manuscript Preparation: Cindy Bonnar
Editing: Jim Coats
Drawings: Marvin Ehrlich, Franz Baumhackl
Design and Production Coordination: Naomi Schiff

Integrated Pest Management for Potatoes in the Western United States

This manual is designed to help growers, pest control advisers, and farm managers apply the principles of integrated pest management (IPM) to potato crops in the western United States. In an IPM program, pest management is coordinated with production practices to achieve economical protection from pest injury while minimizing hazards to crops, human health, and the environment. The emphasis is on anticipating and preventing problems whenever possible. The term *pest* is used in this manual to include destructive insects and other arthropods, pathogens, nematodes, weeds, and vertebrates.

The second chapter of this manual summarizes crop growth and development. It provides a background for understanding how potato growth and production are influenced by pest injury, cultural practices, and other factors. The discussion of management methods in the third chapter places integrated pest management in perspective with other production practices. Later sections on specific pests stress biological information that relates to management strategies and the pest's impact on the crop. The photos and descriptions will help to identify pests and the symptoms of pest damage. Charts and forms are presented to help organize pest management activities. The Glossary includes technical terms and other terminology that may be unfamiliar to some readers.

Pesticides are discussed where appropriate, but because pesticide registrations often change, this manual does not make specific recommendations. However, the references listed at the end of the manual include current recommendation pamphlets from universities in the western states. Research is continuing in many areas of potato pest management, so monitoring and control methods may change as new information and crop cultivars become available. Stay up to date on pest management methods by keeping in touch with local extension agents, farm advisors, or other experts.

The western states (Figures 1 and 2) produce nearly two-thirds of the potatoes grown in the United States on only 50% of the total U. S. potato acreage (Table 1). The majority of acreage is planted to Russet Burbank, which is grown to some extent in all but the low desert valleys. Over 15 other cultivars are also grown in one or more

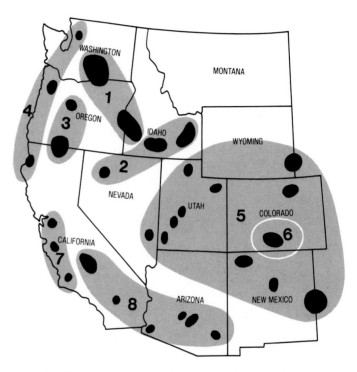

Figure 1. Major production areas for commercial potatoes (potatoes not grown for seed) in the western United States: (1) Columbia Basin and lower Snake River Valley; (2) Upper Snake River Valley and northern Nevada; (3) Central Oregon and Klamath Basin; (4) Northern coastal valleys; (5) Rocky Mountain valleys and plains; (6) San Luis Valley; (7) Sacramento-San Joaquin Delta and central California coastal valleys; and (8) Low-elevation desert valleys.

areas of the region. In most of the growing areas potatoes are planted in the spring and harvested in late summer and fall. An exception is the low desert valley region (area 8, Figure 1), where most of the acreage is planted in fall or winter and harvested in late spring or early summer. The growing season in most areas ranges between 120 and 140 days. However, some high elevation areas have 90- to 100-day seasons, while some areas in the Columbia Basin have 160- to 180-day seasons.

A number of pests are important to all western potato-growing areas. Aphids, primarily the green peach

Figure 2. Major seed potato producing areas of the western United States.

aphid and the potato aphid, are vectors of leafroll and other potato viruses, important diseases wherever potatoes are grown. Verticillium wilt and blackleg can be problems everywhere; bacterial ring rot can occur, but its incidence is minimized by the use of certified seed. Rhizoctonia is always a threat if sprout growth is delayed or occurs in cool, wet soil. In all growing areas nightshades, pigweeds, lambsquarters, and annual grasses are problems, and nutsedge is a potential problem.

Other pests are important only in certain parts of the western region. Root-knot nematodes occur wherever potatoes are grown, but they are most serious in Pacific Northwest growing areas, areas 1 to 3 of Figure 1. Potato psyllid occurs throughout the West; it is a major though sporadic problem in the Rocky Mountain growing areas, areas 5 and 6, and in Arizona. Quackgrass and Canada thistle are problem weeds in most of the Pacific Northwest and Rocky Mountain growing areas, areas 1 to 6. Field bindweed occurs occasionally in all areas; it is more frequently a pest in parts of areas 1,2, and 5. Kochia, cocklebur, and sunflower are common weeds in areas 1,2,3,5, and 6.

Russet Burbank and Norgold Russet are the main cultivars grown in the *Columbia Basin and lower Snake River Valley*, area 1. Other cultivars include Kennebec, Nooksack, Centennial Russet, Lemhi Russet, and Butte. Planting begins in March and April. Early harvests of Norgold Russet begin in July; Russet Burbanks are

harvested from late August through early November. The Columbia Basin of Washington and Oregon has the longest growing season, up to 180 days, with late harvest yields of 500 to 700 hundredweight per acre, the highest reported yields in the world. Yields in the lower Snake River Valley average 300 to 350 hundredweight per acre. Important insect pests in these areas include aphids, Colorado potato beetle, and wireworms. Leafroll and net necrosis, other potato viruses, Verticillium wilt, blackleg, and Rhizoctonia are major disease problems; powdery mildew and Sclerotinia are minor problems. Root-knot nematodes are serious pests. Important weeds include Canada thistle, quackgrass, field bindweed, nutsedge, dodder, nightshades, pigweeds, lambsquarters, kochia, cocklebur, and sunflower.

The vast majority of acreage in the *upper Snake River Valley and northern Nevada*, area 2, is planted to Russet Burbank; other cultivars grown include Norgold Russet, Lemhi Russet, and Butte. Planting takes place in April and harvest in September and October, with yields averaging 250 to 300 hundredweight per acre. Aphids, Colorado potato beetle, wireworms, Verticillium wilt, blackleg, Rhizoctonia, leafroll and other potato viruses, root-knot nematodes, and early blight are important pests in this area. Problem weeds include Canada thistle, quackgrass, nightshades, pigweeds, lambsquarters, kochia, cocklebur, and sunflower.

Central Oregon and the Klamath Basin of southern Oregon and northern California, area 3, produce seed potatoes and Russet Burbank for fresh market. Planting is in May and harvest in September and October, with yields averaging 375 to 400 hundredweight per acre. Frost is a threat both early and late in the season; solid set sprinklers are used to protect the foliage from frost damage. Colorado potato beetle is a pest in central Oregon but not in the Klamath Basin; aphids, wireworms, Verticillium wilt, blackleg, potato viruses, Rhizoctonia, and root-knot nematodes are problems in both areas. Early blight and Phytophthora tuber rot are important pests in the Klamath Basin, where corky ringspot is sometimes a problem in coarse, sandy soils. Some years, meadow mice become serious pests in parts of the Klamath Basin. Major weed pests include Canada thistle, quackgrass, nightshades, kochia, and wild oats. Several mustard family weeds are important, primarily because they can be early season hosts for green peach aphids.

The relatively cool, damp climate of the *northern coastal valleys*, area 4, makes late blight a perennial problem. The possibility of wet weather after planting always makes seed piece decay a potential threat. Chipping cultivars, mainly Kennebec, are planted in northern California in the spring for late summer or early fall harvest. Yields are about 300 hundredweight per acre. Potato acreage in the Willamette Valley of Oregon is planted to chipping cultivars, primarily Norchip, and to

Russet Burbank for fresh market. Planting is in May and harvest from late August to early October. Norchip yields average about 275 hundredweight per acre and Russet Burbank about 375 hundredweight per acre. The main cultivars grown in northwest Washington are Russet Burbank, Norgold Russet, and Norchip. A range of cultivars is grown for seed. Planting is in May and harvest from September through November. Yields of commercial plantings are about 325 to 375 hundredweight per acre. Late blight is the major pest problem; aphids, the viruses they transmit, and Rhizoctonia are also important. Quackgrass and annual grasses are the major weed problems. Because most fields are not irrigated, periods of no rainfall are a problem in these areas.

The *Rocky Mountain valleys and plains*, area 5, have seasons and yields similar to the upper Snake River Valley. Cultivars grown include Russet Burbank, Centennial Russet, Norgold Russet, Norchip, Monona, Sangre, and Red Pontiac. Colorado potato beetle occurs throughout this region, and is a serious pest in the most northern areas. Potato psyllid is a sporadic and occasionally serious problem. Early blight can be a major disease problem in most of the Rocky Mountain areas. Verticillium wilt, blackleg, and potato viruses are also important diseases. Root-knot nematodes are occasionally serious in some areas. Quackgrass is a problem in some areas; Canada thistle is important except in central and northwestern New Mexico. Other problem weeds include nutsedge, field bindweed, nightshades, pigweeds, lambsquarters, and annual grasses.

The *San Luis Valley*, area 6, is at a higher elevation than the other Rocky Mountain valleys. A large proportion of the acreage is planted for seed production. The main cultivar grown for commercial production is Centennial Russet; Russet Burbank, Sangre, and small amounts of several other cultivars are also grown. Planting is in late spring, harvest in early fall, and yields aver-

age about 300 hundredweight per acre. Pest problems are similar to the other Rocky Mountain areas; psyllids are occasionally serious, but Colorado potato beetle is not a problem. Early blight is a serious disease, as well as Verticillium wilt and blackleg. Aphids and viruses transmitted by them are important; however, for an unknown reason, net necrosis does not usually develop in Russet Burbank tubers infected with leafroll virus. Important weed pests include Canada thistle, nutsedge, nightshades, pigweeds, lambsquarters, and annual grasses.

Several potato cultivars are grown in the Sacramento-San Joaquin Delta and central coastal valleys of *central California*, area 7, and yields average 300 to 400 hundredweight per acre. In the Sacramento-San Joaquin Delta, summer crops of primarily White Rose and Red LaSoda are grown, and seed potatoes are planted in the summer for winter harvest. In the Salinas Valley, plantings of Kennebec for chipping are made from March through June for harvest from summer through winter, coinciding with processors' needs. The tubers are frequently stored in the ground for one to three months after vinekill. Russet Burbank is planted in the Santa Maria area in spring for late summer harvest. Tuberworm is a pest throughout central California. As in all other growing areas, aphids and the potato viruses they transmit are important pests. Nematodes generally are not a problem in the peat soils of the Delta; however, the high organic matter content increases the potential for scab. Root-knot nematodes and corky ringspot can be problems in the Salinas Valley. Late blight occurs to some extent every season in both coastal valleys. Blackleg, Rhizoctonia, and Verticillium wilt are always potential problems. Important weed pests in these areas include nightshades, pigweeds, mallow, nettle, annual sowthistle, and mustards.

In the *low-elevation desert valleys* of Arizona and California, area 8, most potatoes are planted from

Table 1. Potato Acreage (Thousands of Acres) and Production (Thousands of Hundredweight [Cwt]) in the Western United States.

	1980		1981		1982		1983		1984		1985	
	Acres	Cwt	Acres	Cwt	Acres	Cwt	Acres	Cwt	Acres	Cwt	Acres	Cwt
Arizona	4.0	1,276	5.2	1,456	4.7	1,434	4.9	1,421	5.4	1,647	5.8	1,450
California	50.5	18,962	56.3	21,071	56.2	21,235	56.2	19,467	60.4	22,767	64.5	22,878
Colorado	42.3	12,545	46.8	13,504	51.4	14,489	49.9	14,764	60.1	19,213	63.0	20,140
Idaho	300.0	79,840	330.0	84,540	339.0	91,710	332.0	86,660	325.0	86,600	355.0	102,515
Montana	6.9	1,725	7.4	1,739	7.4	1,924	7.2	1,944	7.4	1,924	7.2	1,071
Nevada	13.0	4,420	12.0	3,480	13.0	4,095	12.0	3,960	10.0	3,300	9.0	3,105
New Mexico	3.0	540	4.5	945	4.9	1,274	5.7	1,625	9.1	2,639	9.4	2,860
Oregon	47.0	19,745	54.0	21,710	52.5	21,105	48.5	20,710	55.5	22,680	60.0	26,614
Utah	5.2	1,170	5.8	1,276	5.8	1,305	4.9	1,224	6.4	1,728	6.5	1,658
Washington	87.0	49,395	108.0	52,920	110.0	52,800	103.0	53,560	115.0	56,925	126.0	62,370
Wyoming	5.7	1,340	5.3	1,060	5.2	988	5.3	1,060	2.0	520	2.3	245
Total	564.6	190,958	635.3	203,701	650.1	212,359	629.6	206,395	656.3	219,943	708.7	244,906
U.S. Total	1,154.0	302,857	1,273.1	338,591	1,273.5	351,808	1,233.7	292,696	1,300.0	361,648	1,377.0	399,836

November to February and harvested from May to August. The winter weeds nettle and London rocket may be problems in these areas. Black nightshade is the most prevalent nightshade species; mallow, purslane, and Russian thistle are also common weed pests. In California, some plantings are made in the summer and harvested in the winter. White Rose, Centennial Russet, Kennebec, Red LaSoda, Norgold Russet, and Chieftain are the major cultivars grown in California, with yields of 300 to 500 hundredweight per acre. Pest problems include tuberworm, wireworms, scab, Sclerotium stem rot, and occasional outbreaks of late blight. Root-knot nematodes occasionally are serious when tubers are harvested in July or August. In Arizona the main cultivars are Kennebec, Norgold Russet, and Red LaSoda; yields average 280 to 300 hundredweight per acre. Major pest problems include potato psyllid, blackleg, and Pythium tuber rot.

Most seed-producing areas are high elevation, cool, short-season areas (Figure 2). Location, climate, and the certification requirements for fields and seed tubers used for planting help reduce the severity of pest problems in these growing areas. Green peach aphid is the most important pest because it transmits several potato viruses. Control measures are designed to prevent any aphid buildup in seed fields, and in some cases to prevent their development on alternate hosts or in commercial potato fields, from which they can transmit viruses to seed fields. Colorado potato beetle is a problem in Idaho, Wyoming, and some of the areas in Montana and Oregon. Cool weather and the finer-textured soils of many seed areas help keep root-knot nematode populations from building up; certification requirements usually restrict the growth of certified seed potatoes to fields that have been free of root-knot symptoms. Common weed pests include nightshades, pigweeds, lambsquarters, Canada thistle, and quackgrass. Control of mustard family weeds is important because they can be hosts for early season development of green peach aphid populations.

Growth and Development Requirements of the Potato Plant

To accurately evaluate pest injury and distinguish it from other kinds of plant stress, it is important to understand the biology of the crop. IPM methods are chosen for their effects on the crop as a system, not just for their abilities to control specific pests. Successful pest management in potatoes requires that you know how pests and management practices interact to affect potato growth and development.

Computer models of potato growth and development are being developed by university researchers in California, Idaho, Oregon, and Washington. These models use climatic data and certain cultural practice information to "grow" potato crops mathematically, and to predict the effects of management practices and pest damage on yields and quality. At the present time no management recommendations are available based on the crop models; however, in the near future they may prove useful in helping the potato grower make accurate management decisions.

Growth Requirements

The potato plant, like a factory, requires a supply of raw materials and a source of energy to make a product. The raw materials for plant growth and tuber production are carbon dioxide, oxygen, water, and soluble mineral nutrients. Like other green plants, potatoes use complex chemical structures including chlorophyll, the green pigment of plants, to capture energy from sunlight. Some of the trapped energy is used to convert water and carbon dioxide from the air into sugars and other compounds that serve as the plant's energy supply and as building blocks for growth. This process is called *photosynthesis*.

Most water absorbed from the soil by plants passes from the roots up through the vascular system and evaporates through leaf pores called stomata. Only a small amount of the water taken up by plants actually remains in their tissue. Most evaporation from leaves, which is called *transpiration*, occurs because stomata remain open to expose a moist surface that can absorb carbon dioxide from the air. The flow of water from roots to leaves also supplies water used in photosynthesis, carries nutrients throughout the plant, and serves as a means of cooling. Higher temperature, lower humidity, and stronger wind generally increase transpiration and, therefore, the amount of soil moisture withdrawn by plants. However, under extremely hot, dry conditions, stomata will close, stopping transpiration.

Plants absorb soluble mineral nutrients from the soil to build proteins, chlorophyll, and other components. Nutrients needed in the largest amounts are nitrogen, phosphorus, potassium, calcium, magnesium, and sulfur. Nutrients required only in very small amounts but still essential are iron, boron, manganese, zinc, molybdenum, copper, and chlorine.

Respiration is the process by which sugars and other compounds are metabolized to provide energy used to drive many growth processes. The availability of oxygen is critical for the respiration of roots and tubers as well as for aboveground plant parts. Both roots and tubers become more susceptible to certain diseases or physiological disorders in waterlogged or poorly aerated soil.

Development

In nature, the potato is a perennial plant that survives from year to year as a tuber, which is a modified underground stem. When grown as an annual crop, potatoes are propagated vegetatively through the planting of tubers or parts of tubers (*seed pieces*) rather than true seeds. Potato plants do produce true seeds, but in the United States these are used for propagation primarily by plant breeders developing new cultivars.

A number of diseases can be transmitted in potato tubers. Potato plants that grow from tubers left in fields or cull piles are important sources of inoculum for these diseases. Foundation seed and seed certification programs have been developed to obtain and maintain sources of seed tubers that are relatively free from disease.

Growth and development of potato plants can be divided into four stages (Figure 3):

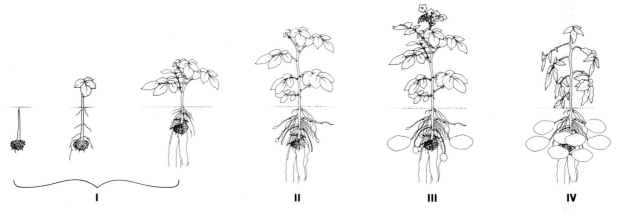

Figure 3. Growth stages of the potato plant. I *Vegetative Growth*: sprouts emerge and form first 8 to 12 leaves. II *Tuber Initiation*: tubers begin to form at tips of stolons; foliage continues to develop.

III *Tuber Growth*: most of plant production supports tuber enlargement. IV *Maturation*: tuber periderm (skin) thickens, and dry matter content reaches a maximum; vines begin to senesce.

- Vegetative Growth (stage I)
- Tuber Initiation (stage II)
- Tuber Growth (stage III)
- Maturation (stage IV)

The timing of the events that make up this pattern of growth is influenced by cultivar, age of seed pieces, weather, cultural practices, pest damage, and other factors.

Vegetative Growth

The potato tuber is a modified stem and its *eyes* are clusters of modified buds (Figure 4). Either whole tubers or pieces of tubers, each with one or more eyes, are planted. The emergence of sprouts from eyes is favored by dark, warm, moist conditions. Sprouts grow upward to emerge from the soil and form the stems of the plant. Buds form at swollen nodes along the stem. Leaves and branch stems develop from aboveground nodes. Roots and underground stems called stolons develop from underground nodes. Tubers develop on the ends of stolons.

Before emergence, seed pieces may decay, and developing stems are susceptible to certain diseases, such as Rhizoctonia and blackleg. Vigorous sprout growth is more resistant to infection. Generally, the more rapidly sprouts emerge, the less susceptible they are to damage. The rate of sprout growth and, consequently, the time until emergence are temperature dependent (Table 2).

Starch in the seed piece supplies energy for sprout growth and development. There is usually enough starch in a sound potato seed piece to support plant growth for 30 days. About half of it is used up by the time the plant starts to produce sufficient energy to support further growth and development and accumulate reserves for tuber production. Food reserves and mineral nutrients continue to be used until they are exhausted or the seed piece is decayed. Inorganic nutrients, some of which must be supplied as preplant fertilizer, are generally necessary for most active early growth.

Roots develop from the base of each sprout, from underground nodes, and sometimes from stolons and tubers (Figure 4). The root system that develops is highly branched. Most of the nutrients that move up through the vascular system (Figure 5) are absorbed by fine root hairs at the tips of root branches. By the time sprouts emerge, there is an extensive root system above the level of the seed piece. A fully developed plant has a root system that is concentrated in the top 18 inches (45 cm) of soil, although in some cases roots may extend as deep as 5 feet (150 cm).

Usually one to three sprouts from each seed piece develop into the mainstems of the plant. Leaves develop after the stems emerge from the soil. The potato leaf is compound, consisting of a petiole with a terminal leaflet, two to four pairs of primary leaflets, and secondary leaflets interspersed along the petiole (Figure 4). Leaflet

Table 2. Approximate Times for Russet Burbank Sprouts to Emerge, When Planted at Different Soil Temperatures and Depths. (Emergence times are affected by cultivar, conditions of seed storage, and soil physical condition.)

Soil Temperature at Planting (°F)	Days to Sprout Emergence at Planting Depth	
	4 Inches	6 Inches
40°	40+	40+
45°	30	40
50°	25	28
55°	22	26
60°	20	24
65°	18	22

characteristics depend on cultivar. Once leaves are formed, a potato plant is able to produce all the energy it needs for further growth and development.

Stolons are lateral shoots that develop from buds at the underground nodes of stems. The vascular system of a stolon has a large amount of phloem tissue for rapid translocation of nutrients into growing tubers (Figure 5). Tubers develop just behind the tips of stolons. If stolons emerge from the soil, they form leafy shoots called secondary stems. Stolons produced on secondary stems do not always form tubers.

Depending on their growth habit (pattern of vine growth), potato cultivars may be classified as determinate or indeterminate. Late season cultivars such as Russet Burbank and Centennial Russet are *indeterminate*. They have a vine type of growth habit, forming a succession of branch stems that become the main growing point and produce new foliage and flowers. Production of the first flowers (*primary flowering*) approximately coincides with the beginning of the tuber growth phase and occurs after the mainstem has formed 8 to 12 nodes. A branch then develops from an axillary bud one or two nodes below the flowering tip, producing 8 to 12 nodes and then secondary flowers. After *secondary flowering*, another branch develops in a similar fashion from below the secondary flowers; its growth terminates with a *tertiary flowering*. Indeterminate cultivars continue producing new foliage and in some cases may keep initiating new tubers until disease, weather conditions, or diminishing nutrients or water stop their growth. Completion of their growth cycle requires 100 or more days.

Early season cultivars such as Norgold Russet and Norland have a *determinate*, bush type of growth habit. Branches may develop after primary flowering, but always from nodes on the main stem. Tuber initiation and the onset of tuber growth occur earlier in determinate cultivars, which do not continue to produce new foliage throughout the season. Determinate cultivars complete their growth cycle in 60 to 80 days, even if nutrient availability and environment continue to be favorable. These cultivars may exhibit indeterminate type growth if grown under warm night conditions, when minimum temperatures do not fall below 60° to 65°F (15° to 18°C).

Tuber Initiation

The onset of tuber initiation (*tuberization*) is controlled by growth-regulating hormones produced in the potato plant. Before tuber initiation can start, more carbohydrate must be produced by photosynthesis than is needed to support the growth of leaves, stems, and roots. The excess carbohydrate, in the form of sucrose, is translocated down to stolon tips, where tuber initiation begins. Tuber initiation usually begins when the leaf area

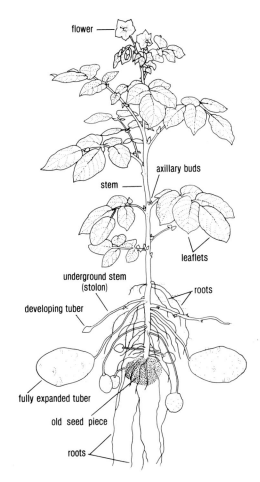

Figure 4. Diagram of a potato plant. (Adapted from *Commercial Potato Production in North America, Potato Association of America Handbook,* Potato Association of America, Orono, ME, 1980, 36pp.)

index (the ratio of total plant leaf area to the area of soil surface covered by the plant) is 1.5 to 2.0. In indeterminate cultivars such as Russet Burbank this occurs at the 8- to 12-leaf stage; determinate cultivars such as Norgold Russet may begin to initiate tubers earlier. Tuber initiation is affected by soil moisture, soil temperature, and nitrogen management.

Tubers usually first appear on the lower, older stolons, when high amounts of sucrose begin to build up in the stolon tips. During tuber initiation the area just behind the stolon tip undergoes cell division and many of the axillary buds become eyes. Sucrose can be converted into starch for storage in the expanding tuber cells before tuber initiation is visible. Under normal conditions, Russet Burbank tubers that will subsequently reach harvestable size are initiated over a 10- to 14-day period and the tuber initiation phase is completed by

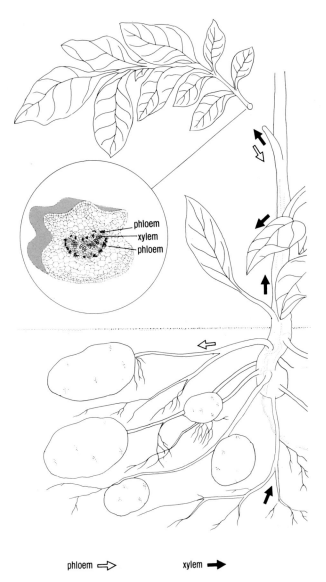

phloem ⇨   xylem ➡

Figure 5. Diagram of the conducting tissues (vascular system) of a potato plant. The xylem carries water and nutrients from the roots, and the phloem distributes the products of photosynthesis.

about the 12-leaf stage. Indeterminate cultivars may continue to initiate tubers after the end of this growth phase, but these late tubers may not reach harvestable size.

Tuber Growth

Although some cell division occurs, tuber growth is primarily a period when cells formed during the initiation phase expand 7- to 18-fold with water and starch reserves. Tubers become the dominant sink for the carbohydrates produced by photosynthesis and for inorganic nutrients such as nitrogen, phosphorus, and potas-

sium. Carbohydrates move into the expanding tubers primarily as sucrose, which is rapidly converted into starch. The microscopic starch grains formed make up 10% to 25% of the fresh weight of a tuber. As tuber cells accumulate starch, the ratio of dry matter content to water content increases. Tuber dry matter content is measured as *specific gravity*; the higher the dry matter, the higher the specific gravity. Higher specific gravity is more desirable, especially in tubers to be used for processing.

If disease, inadequate fertilization, or other factors reduce the ability of the plant to take up inorganic nutrients, nutrients are drawn from the vines to support continued tuber growth. The vines then begin to show symptoms of aging (senescence) prematurely. They become more susceptible to diseases such as Verticillium wilt, early blight, and blackleg.

Because tubers are modified stems, they have the same kinds of structures as aboveground stems (Figure 6). Each eye of a tuber contains several axillary buds that can give rise to new stems. Normally one bud is dominant and is the only one that will sprout. However, if this sprout is removed, other buds within the eye will sprout.

The *periderm* (skin) of a tuber is made up of a single layer of corky cells called the epidermis and several layers of sloughed-off corky cells above the epidermis. When tubers are injured, by bruising or cutting for example, a wound periderm forms. It consists of several layers of corky cells containing a high proportion of a waxy substance called suberin; the formation of wound periderm is called *suberization*. Suberin contains compounds toxic to microorganisms and acts as a barrier to invasion by disease organisms. Suberization is important in protecting stored tubers and seed pieces from infection.

Lenticels are surface pores present on both stems and tubers (Figure 6). They are the same kind of structure as leaf stomata and, like stomata, they serve as sites for gas exchange. Layers of suberin are present underneath the lenticels and normally act to prevent infection. However, under conditions of high moisture, cells in the underlying cortex layer may swell and break through the suberin layers causing open or *enlarged lenticels*. Enlarged lenticels are susceptible to disease infection.

The tuber has a vascular system, part of which is concentrated in an area called the vascular ring and part dispersed throughout the tuber. The tuber xylem carries water and the phloem carries carbohydrates and other nutrients to expanding cells. Certain physiological conditions or disease organisms can cause portions of the tuber vascular system to turn dark brown. Tuber tissue lying between the vascular ring and the periderm is called the cortex; tissue inside the vascular ring is called the medulla. A watery region called the pith is present in the center of the medulla, and extends outward to the eyes.

Maturation

During the maturation period, tuber skin (periderm) thickens or *sets*, tuber dry matter content reaches a maximum, and vines senesce. Senescing vines begin to turn yellow and lose leaves, and photosynthesis decreases. Increases in tuber dry weight during this period are mainly due to movement of carbohydrates from vines into tubers. During this time eyes become dormant, and starch synthesis decreases and then finally stops when vines die. The periderm continues to thicken after vine death.

Energy Resources and Yield

Producing a high yield of quality tubers at reasonable cost requires the right balance between the *supply* of energy produced by the plant and the *demand* placed on this energy supply. The energy supply for a potato plant consists of carbohydrates produced by photosynthesis. Carbohydrates supply energy needed for two basic functions: respiration and growth.

Respiration, the process that supplies energy needed to support the metabolism of living cells, always has first priority in its demand for energy. The energy supply remaining after respiration needs are met is available for tissue expansion and deposition as reserves. However, plants cannot support the growth of roots, stems, leaves, and tubers at uniform rates simultaneously. Thus, the energy supply is allocated according to a system of priorities (Figure 7).

During the vegetative phase, growth of roots, stolons, stems, and leaves has priority for energy after respiration. The plant builds the vegetative framework needed to synthesize and distribute carbohydrates necessary for tuber growth. Once its leaf area index reaches 3 to 4, a potato plant is able to produce all the energy it needs to sustain maximum tuber growth.

When the foliage accumulates more carbohydrate than is needed to support vine growth, the surplus is available for tuber initiation and growth. As more and more tubers are initiated and begin to expand, tuber growth takes precedence over vegetative growth and initiation of new tubers. By far the greatest amount of tissue production occurs in tubers (Figure 8). The rate of tuber growth depends on the availability of carbohydrate in excess of that required for foliar growth and plant metabolism. The more carbohydrate produced by the vines, the faster tubers grow; maximum tuber growth is supported as long as the leaf area index is above 3.0 and conditions are favorable.

If excess nitrogen is applied to indeterminate cultivars at planting, excess foliar growth occurs and the

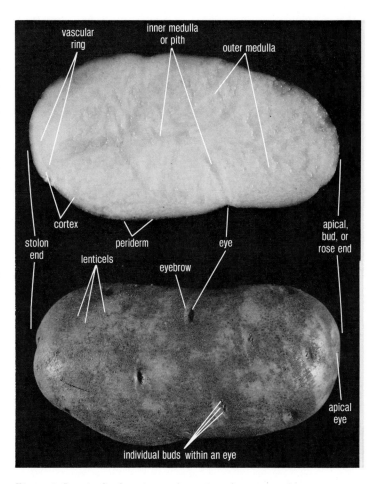

Figure 6. Longitudinal section and exterior of a potato tuber, showing structural features. Xylem and phloem are located in the vascular ring and throughout the medulla tissue. After the dormancy period eyes gain the ability to form sprouts, starting at the apical eye and proceeding from the bud to the stolon end.

onset of tuber growth may be delayed by as many as 10 to 14 days. Although maximum leaf area index increases, tuber growth rate does not increase. Therefore, the delayed tuber growth may result in reduced yields if the season is too short to allow completion of the growth cycle. Application of excess nitrogen to determinate cultivars results in smaller increases of leaf area index and does not delay the onset of tuber growth.

Effects of Pest Injury on Yield

Pest injury often affects the production, transport, or allocation of energy resources within the potato plant. When pests interfere with these important plant functions, they may severely limit yields. Weeds may reduce yields by competing with potato plants for available water and nutrients.

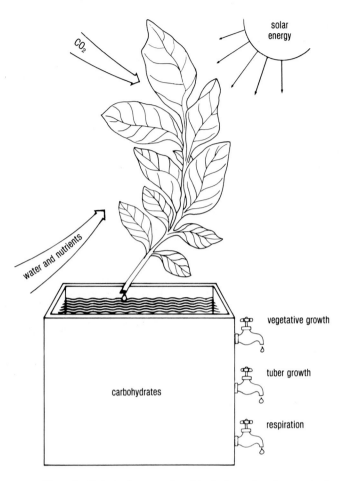

Figure 7. Carbohydrates produced in the leaves by photosynthesis are an energy supply that supports plant functions. As demand drains the pool of energy, vegetative growth ceases first, then tuber growth, and finally respiration.

Water and Nutrient Transport. Nutrients and water move from plant roots to the leaves and stems through xylem tissue within the vascular system. Certain pathogens restrict flow within the xylem, interfering with water and nutrient transport and causing wilting and premature aging of potato vines. Verticillium wilt and blackleg are the most important diseases causing xylem blockage. Ring rot and severe *Rhizoctonia* infections are also very effective in reducing xylem transport.

Carbohydrate Transport. A number of diseases may reduce or block the transport of carbohydrates from foliage to tubers by clogging the phloem tissues of the vascular system. Small, green tubers called *aerial tubers* form at the base of a stem when sugar accumulates because it cannot be transported into tubers fast enough. Sucrose also accumulates in leaves, causing them to thicken and become brittle and papery. Leaves and stems of red-skinned varieties turn reddish. Among the pests that can cause these symptoms are potato leafroll virus (aerial tubers are rarely caused by leafroll virus), blackleg, *Rhizoctonia*, curly top virus, aster yellows, and witches' broom. Conditions that reduce movement within the phloem, such as waterlogging of the roots, can also cause similar symptoms.

Foliage Reduction. Some pests affect crop production by reducing the leaf area available to intercept sunlight and perform photosynthesis. Because fully developed potato vines have more leaves than required for maximum tuber growth, vines can sustain some loss in leaf area before a reduction in photosynthesis limits tuber growth. However, foliage losses that do not reduce photosynthetic output may cause the plant to accelerate vine growth at the expense of tuber growth, delaying or reducing yields.

The stage of growth when foliar damage occurs is critical. If foliage loss occurs before adequate leaf area is achieved, it may be difficult for plants to recover; tuber production may be delayed or prevented. Damage that occurs during tuber initiation or growth causes a greater loss in yield than damage occurring during the maturation phase. For example, early blight often occurs late in the season, when the loss of foliage for photosynthesis does not result in significant yield reduction. However, if early blight develops soon enough, controls are needed to prevent yield loss. Foliage feeding by Colorado potato beetles is also much more serious when it occurs early in the season.

Second Growth

If water uptake is less than water loss from transpiration during hot weather, water stress in leaves will cause stomata (leaf pores) to close. Photosynthesis then declines as carbon dioxide availability is reduced by stomatal closure. This interrupts the flow of carbohydrate that maintains normal tuber growth. When conditions again become favorable for photosynthesis and carbohydrate levels increase, improving the supply to tubers, growth of individual tubers may resume unevenly. Some areas of a tuber may grow more slowly than others; some areas may not resume growth. Uneven tuber growth results in various types of tuber malformation, called second growth disorders. The parts of tubers that grow slowly or do not resume growth may have lower starch concentrations and higher sugar levels. Such tubers are unacceptable for processing. Cultivars vary in their susceptibility to second growth disorders; Russet Burbank is highly susceptible.

Storage

Tubers continue to respire after harvest. Energy derived from respiration is required to support suberization during the curing period and to support metabolic processes that continue in storage. Adequate storage ventilation is necessary to provide oxygen for respiration and to control humidity.

Sugar Accumulation.

When tubers are stored at temperatures lower than 45° to 48°F (7° to 9°C), depending on the cultivar, some starch will be converted into sugars. The lower the temperature, the faster this conversion occurs. Sugar accumulation is undesirable in tubers used for processing because the sugars caramelize during processing, darkening the product. For this reason, tubers to be used for chips usually are stored at no less than 50°F (10°C), and tubers used for french fries no less than 45°F (7°C). If sugars do accumulate in storage, their levels may be reduced by holding tubers at temperatures above 50°F (10°C), a process called *reconditioning*. Some varieties are less likely to accumulate sugars or are more easily reconditioned, making them more desirable for processing.

Dormancy

After potato tubers are harvested, a certain length of time is required before the eyes can sprout. This time is called the *dormancy* period; its length depends on cultivar and conditions during growth and storage. Hot weather during growth and high or fluctuating storage temperatures shorten the dormancy period. At the end of the dormancy period the apical eye at the bud (*rose*) end of the tuber is dominant over the others; it is the only one that will sprout unless the tuber is cut or the apical sprout is removed. A *pical dominance* weakens with time

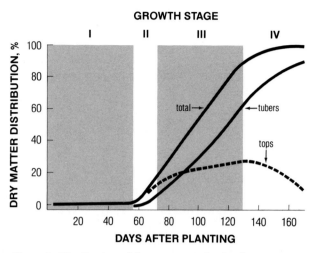

Figure 8. Distribution of dry matter production between tops (vines) and tubers of Russet Burbank potato plants. These curves are characteristic of indeterminate cultivars. Determinate cultivars such as Norgold Russet have similar curves, but the growth stages begin earlier. The actual timing of growth cycle events depends on local growing conditions.

after dormancy is over, until eventually all eyes will sprout. Seed tubers are generally stored at 35° to 38°F (1.7° to 3.3°C) to delay sprouting.

Aging

The respiration that takes place during storage slowly uses up some of the starch reserves that are needed to support development of sprouts from planted seed tubers. Other metabolic changes take place as well during storage, including the loss of dormancy and apical dominance. The net effect of these changes is an aging of tubers that is sometimes called *physiological aging*. Some aging of seed tubers is desirable to achieve proper sprouting of seed pieces. Older seed tubers produce more stems per hill and set more tubers. However, if seed tubers are too old, weak plants may be produced or little tuber disorders may result.

Managing Pests in Potatoes

No single management program is suitable for all potato crops. Pest problems vary from field to field and season to season because of differences in soil type, cropping history, cultural practices, cultivar, and the nature of surrounding land. Choice of market and market conditions also affect the feasibility of management options because they determine how a crop must be handled and the value of that crop. Regardless of conditions, however, four components are essential to any IPM program:

- accurate pest identification;
- field monitoring;
- control action guidelines;
- effective management methods.

The purpose of this chapter is to help you incorporate these components in a program tailored to the specific needs of each field. Management practices that are important in controlling potato pests are summarized in Table 3.

Pest Identification

Because most pest management tools, including pesticides, are effective only against certain pest species, you must know which pests are present and which are likely to appear. Different control methods may be needed even for closely related species.

Use the descriptions and photos in this manual to identify pests commonly affecting potatoes in the western United States. Check the References for other sources of information. *Remember that the help of experienced professionals is required to reliably diagnose some pest problems; don't hestiate to seek technical help if you are not sure of an identification.* Extension agents, farm advisors, or other experts in your area can help you identify pests and can direct you to other specialists if necessary.

Monitoring

By monitoring your field, you can get the information you need to make management decisions. Monitoring in-

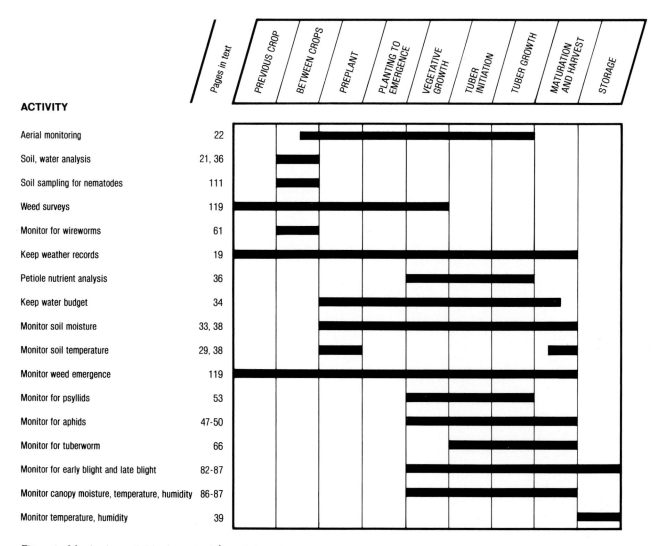

Figure 9. Monitoring activities important for potato pest management.

cludes keeping records of weather, crop development, and management practices, as well as evaluating incidence and levels of pest infestations. Seasonal monitoring activities are summarized in Figure 9.

Weather Data

Because weather influences the development of the potato plant and its pests, a reliable source of weather data will improve your pest management decisions. Daily high and low temperatures are needed if degree-day accumulations are used to schedule early blight control. Evapotranspiration data may be useful for scheduling irrigations, and weather forecasts are essential for scheduling planting, harvest, frost protection irrigation, and other operations. Many newspapers and radio stations in agricultural areas report such information; the National

Weather Service broadcasts weather information on NOAA Weather Radio, VHF channels at 162.42, 162.50, and 162.55 MHz.

Regional weather information may not reflect the situation in your field, which is affected by local variations in terrain, vegetation, elevation, and other conditions. The best source of temperature data is a recording maximum/minimum thermometer or similar instrument placed as near as possible to the field you are managing. Records of temperature, relative humidity, and rainfall are useful for scheduling late blight sprays in the central and eastern United States. Procedures used there are being studied for adaptation to conditions in the western states; as yet, no specific recommendations are available. Follow the manufacturer's recommendations in setting up weather instruments; in general, thermometers should be 5 feet (1.5 m) above the ground, sheltered from the

Table 3. Summary of Management Activities Important for Controlling Potato Pests.

PEST COMPLEX	MANAGEMENT ACTIVITIES	PEST COMPLEX	MANAGEMENT ACTIVITIES
● Previous Crop		**● Preplant**	
Seed tubers	Visit seed production areas to observe seed fields. Mature seed crop with use of vine killing agents or vine removal if necessary. Harvest seed tubers with minimum mechanical damage. Harvest seed tubers before exposure to freezing temperatures.	Insects and related pests	Control weed hosts. Use soil-applied insecticides when necessary.
		Diseases	Use highest quality certified seed tubers. Suberize cut seed or plant immediately after cutting. Use seed treatment fungicides when necessary. Provide proper planting conditions: • Correct soil structure • Soil temperature above 45°F (7°C) Establish soil moisture that will remain adequate until emergence.
Insects and related pests	Destroy weed host sources. Use rotations that suppress potato insects.		
Diseases	Destroy weed hosts. Destroy potato volunteers. Use rotations that suppress potato pathogens.		
Nematodes	Use rotations that suppress problem species of nematodes.	Physiological disorders	Store seed tubers and cut seed properly. Obtain desirable seed piece size. Provide proper planting conditions.
Weeds	Use rotations that allow control of problem weeds. Avoid herbicides that leave residues harmful to potatoes the following year.		
		● At Planting	
		Insects and related pests	Use systemic insecticides properly when necessary.
● Between Crops		Diseases	Use good sanitation during seed cutting and planting. Sample and monitor for diseased seed pieces. Plant at correct depth in warm (above 45°F [7°C]), moist soil.
Seed tubers	Store seed tubers in proper environment.		
Insects and related pests	Destroy weed host sources. Monitor for wireworms.		
Diseases	Eliminate cull piles. Destroy potato volunteers. Use soil fumigation when necessary and feasible.	Physiological disorders	Provide proper soil fertility. Plant at correct depth and spacing.
Physiological disorders	Analyze for soil fertility, chemistry, and physical constraints.	Nematodes	Use nematicides if necessary.
Nematodes	Use tillage to destroy crop residues. Analyze soil for nematode species. Use fall fumigation when necessary and feasible.	**● Preemergence**	
		Diseases	Avoid preemergence irrigations. Plant shallowly to hasten emergence; form hills after emergence.
Weeds	Use tillage and herbicides to control perennials and potato volunteers. Allow breakdown of harmful herbicide residues.	Weeds	Apply preemergence herbicides when necessary.
		● Emergence Through Tuber Growth	
		Insects and related pests	Monitor for insects and mites. Destroy volunteer host sources. Hill to prevent exposure of tubers to pests. Apply pesticides when necessary.
		Diseases	Maintain proper soil moisture and fertility. Avoid extended periods of wet foliage. Monitor for disease development. Apply fungicides when necessary.

PEST COMPLEX	MANAGEMENT ACTIVITIES
Physiological disorders	Maintain proper soil moisture and fertility. Hill to prevent exposure of tubers to heat and light.
Weeds	Monitor for weed emergence. Time hilling to control emerging weeds. Apply postemergence herbicides when necessary.

● **Maturation and Harvest**

Insects and related pests	Keep tubers covered to avoid tuberworm damage. Use early vinekill to avoid aphids and frost.
Diseases	Maintain proper soil moisture. Apply foliar fungicides or vine killing agents when necessary for late blight control. Harvest after complete vine death to reduce late blight infections. Use careful harvesting and handling procedures. Harvest at proper soil moisture and temperature. Maximize soil removal before storage.
Nematodes	Time harvest to avoid tuber damage when possible. Isolate lots with infested tubers.
Weeds	Apply contact herbicides for complete vinekill and control of nutsedge. Manage water for late-season weed control.

● **Storage**

Diseases	Use proper sanitation in storage areas. Use proper curing conditions. Use ventilation to hasten drying of wet tubers. Maintain proper ventilation, humidity, and temperature. Monitor storages for disease development.
Physiological disorders	Maintain proper ventilation, humidity, and temperature.
Nematodes	Monitor tubers going into and coming out of storage for symptoms of nematode damage. Store at temperatures that prevent nematode multiplication if compatible with intended use.

sun and rain, and placed away from paved surfaces that radiate heat. Recording instruments for late blight forecasting must be placed within the potato canopy.

If you are using degree-days to forecast early blight, you can use tables (see Table 12, page 83) or you can make your own calculations (see Figure 34, page 84). Methods for calculating degree days are discussed in *Degree-Days: The Calculation and Use of Heat Units in Pest Management*, listed in the References.

Pest Activity

Monitoring during the growing season is devoted mostly to checking for insect populations and appearance of diseases. Certain problem weeds such as dodder may require frequent monitoring in specific areas. Visit each field at least once a week; during critical periods of crop development or when a population is approaching a treatment threshold, visit the field every 2 or 3 days. In each growing area, the emphasis is usually on a few pest species that occur regularly and threaten yield or quality. Work your sampling methods into a single routine so that you do not have to make a special pass through the field for each one. In seed potato fields it is important to keep foot traffic to a minimum to reduce the spread of diseases that are mechanically transmitted. Always disinfect footgear before entering seed potato fields.

Other pest monitoring activities are needed only once or a few times each season. For instance, conduct a weed survey late in the season of the previous crop, and repeat the survey once or twice during the growing season and again at harvest. In areas where nematodes are a problem, collect soil samples for laboratory extraction of root-knot and other nematodes in the fall before potatoes are planted.

Soil, Water, and Tissue Testing

Contact a reliable laboratory to test soil, water, and plant tissue. Preplant soil analysis helps determine fertilizer requirements. Soil tests can also measure salinity, soil reaction (pH), organic matter content, and other factors that affect crop growth and management practices.

Occasional testing of irrigation water helps detect harmful increases in salinity, especially where the water comes from wells or where surface water is known to be high in salts. Changes in cultural practices may be necessary to reduce salt problems. Tests should include sodium and boron levels and the concentration of total salts. They should also measure the level of nitrates in the irrigation water, information valuable for planning your fertility program, particularly in the western states where the nitrate level in many wells is increasing.

Analysis of plant tissue is the best way to monitor the crop nutrient status during the growing season.

PLANT GENETICS, INC.

Figure 10. In the first stages of certified seed potato production, tissue culture techniques are used to produce disease-free potato plantlets and microtubers.

Taking at least three or four petiole samples during the period between tuber initiation and early maturation will give you considerable information about the health of the crop and fertility deficiencies, excesses, or imbalances. Results are useful for planning the next season's preplant fertilizer applications, and early analyses can assist in current-season nitrogen management.

Aerial Monitoring

Aerial infrared (IR) photography may be useful for soil mapping and quickly assessing the status of general plant health over large areas. It can detect areas in the field stressed by factors such as inadequate moisture, poor soil fertility, insects, or diseases, and indicate where monitoring on the ground is necessary. If this service is available in your area, check with your Extension Service and with the operator who provides the service to see what this type of monitoring may do for you. Make preseason arrangements to have your fields photographed and to have analyses of the photographs provided regularly.

Keeping Records

Knowing what happened last year can make many management decisions easier. Keep a file, notebook, or computer record of the following for each field:

- weekly monitoring reports;
- weed surveys;
- records of fertilizer and pesticide applications, including names of materials, rates, dates, and methods of application;
- laboratory reports;
- aerial photographs with reports attached;
- agronomic information, including crops and cultivars planted, planting and harvest dates, yields, and grade out;
- disease reports.

In addition to maintaining records that apply to specific fields, keep a file of local weather data, including charts of accumulated degree-days if you are using them to make pest control decisions.

Control Action Guidelines

Control action guidelines indicate when management actions, including pesticide applications, are needed to avoid losses due to pests or other stresses. Guidelines for some insect pests are numerical thresholds based on specific sampling techniques; they are intended to reflect the population level that will cause economic damage if left unchecked. Guidelines for other pests, including most pathogens and weeds, are usually based on the history of a field or region, the stage of crop development, observed symptoms or damage, weather conditions, and other observations.

Control action guidelines often change as new cultivars and cultural practices are introduced and as new information on pests becomes available. The guidelines presented in this manual may be revised as more research is completed on potato pest management.

Management Methods

The ideal IPM program protects the crop while interfering as little as possible with long-term maintenance of the production system. The least costly, most reliable way to deal with pest problems is to anticipate and avoid them. When pesticides are needed, choose materials and application methods that control pests effectively with a minimum of harmful side effects.

Seed Quality and Seed Certification

A number of pests can be transmitted in infected seed tubers, including potato viruses, bacterial ring rot, blackleg, late blight, scab, wilt diseases, and root-knot nematodes. Stem cutting and micropropagation techniques have been developed to obtain pest-free potato plants for propagation and production of certified seed tubers. Disease-free stem cuttings or tiny pieces of meristem tissue are cultured and propagated under sterile conditions to produce large numbers of disease-free plantlets or minitubers (Figures 10 and 11). Several generations of plants are grown in the field to produce certified seed tubers that will be sold to commercial growers.

Certified seed tubers are not guaranteed to be disease free. They are certified to have shown no more than certain low percentages of pest and disorder symptoms during the inspections required by a state's seed certification program. The allowable level of symptom expression for each pest or disorder is called a tolerance level, and these levels vary from state to state. A zero tolerance exists for certain pests, such as bacterial ring rot and root-knot nematode. To pass these tolerances, seed lots must be inspected at least twice in the field during the growing season, and inspected in storage or at the time of shipment. Samples of each seed lot are grown and inspected in winter field trials in Oceanside, California, or in winter greenhouse tests. Pests for which tolerances are enforced and the range of those tolerances among the western states are listed in Table 4. For more detailed information about seed certification programs in individual states, consult the References.

Extra precautions are taken to reduce the incidence and spread of pests in fields where seed potatoes are grown. Most seed potatoes are grown in cool, short-

PLANT GENETICS, INC.

Figure 11. Small tubers produced on plantlets grown under greenhouse conditions may be used to grow the first generation of seed potatoes in the field.

Table 4. Range of Tolerance Levels for Generation IV Plants in Western States Seed Certification Programs.

Disease Symptom	Range of Tolerance Level (%)
Leafroll	0.05–0.20
Potato virus X	4.00–6.00
Mosaic	0.2 –2.0
Calico	1.00[a]
Total all other viruses	0.50–2.0
Haywire	1.00[a]
Witches' broom	1.00[a]
Spindle tuber	0.00–0.25
Blackleg	0.25–5.0
Bacterial ring rot	0.00
Eumartii wilt	0.00
Wilts	2.00
Late blight	0.00[b]
Giant hill	0.50[a]
Root-knot nematode	0.00
Varietal mix	0.05–0.5

a. Montana
b. Colorado

season areas (see Figure 2) where pest populations, including vectors of potato viruses, remain low and the symptoms of virus diseases are often more pronounced, making identification of infected plants easier. However, ring rot symptoms are less pronounced under these same cool conditions. Ideally, seed fields are isolated from commercial fields and home gardens, from which potato viruses could be transmitted by aphids.

When growing seed potatoes, clean all equipment thoroughly before entering fields and keep mechanical contact to a minimum to reduce spread of certain pathogens. For early generations, leave every third row empty to reduce contact during inspections. Minimize cultivation, avoid cultivation when plants are tall enough to be hit by equipment, use herbicides instead of cultivation to control weeds, use solid set or self-moving sprinklers if possible, and keep people and animals out of fields as much as possible. Whenever work in a field is required, do it when the vines are dry; mechanical transmission of pathogens occurs more readily when vines are wet. Remove diseased plants, including any tubers that are present, as early in the season as possible. If adjacent plants are touching, diseased plants should be removed when they are dry and immediately placed in plastic or other smooth, nonabrasive bags to keep them from touching other plants. Fields with a history of ring rot, root-knot nematode, or certain other problems may not be used for certified seed tuber production in some states until nonhost crops have been grown for a specified number of years. Sanitize shoes and equipment between use in different fields or with different seedlots.

Aphid control programs are designed not only to prevent the development of aphid populations in seed fields, but also to prevent populations from building up elsewhere and moving into seed fields. Communitywide programs may be necessary in some seed-growing areas to prevent buildup of aphid vectors in commercial fields or gardens, or on alternate hosts. In some areas systemic insecticides are applied at planting or, where seasons are longer, when about 75% of the plants have emerged. Insecticide sprays are used if aphids appear in seed fields. Controlling mustards and other weed hosts of aphid vectors in field borders and adjacent areas helps prevent the development of potentially damaging aphid populations.

Be familiar with the certification requirements that apply to the seed tubers you buy. Each state uses colored tags to identify quality of seed tuber lots, and printing on the tags identifies generation; learn what these tags mean. Obtain seed tubers from a source with a reputation for supplying good quality seed. If possible, visit the areas where the seed is grown to investigate growing conditions and cultural practices that influence seed quality. Follow careful handling and sanitation procedures to keep certified seed tubers from becoming contaminated. Holding seed tubers under proper conditions for suberization and keeping seed-cutting equipment clean are worthwhile precautions to protect your investment in good-quality seed tubers, as well as the time and expense invested in growing the crop.

Biological Control

Any activity of a parasite, predator, or pathogen that keeps a pest population lower than it would be otherwise is considered biological control. One of the first assessments that should be made in an IPM program is the potential role of natural enemies in controlling pests. Control by natural enemies is inexpensive, effective, self-perpetuating, and not disruptive of natural balances in the crop ecosystem.

The use of insecticides in potatoes reduces the effectiveness of many natural enemies in potato fields. Because of certification requirements and the potential for virus spread into potato fields, damage thresholds in seed potato fields and in commercial Russet Burbank fields in many areas are so low that the activity of natural enemies would not keep insect populations sufficiently controlled. However, natural predators in field borders or adjacent wild areas where insecticides are not acting may serve to keep reservoirs of certain pests such as aphids and Colorado potato beetle at lower levels.

To gain the greatest benefit from biological control of insect pests whenever possible, choose insecticides, rates, and application methods that have a minimal impact on pests' natural enemies. Natural enemies that affect nematodes, weeds, and fungi are being studied, but as yet no practices are recommended for improving biological control of these pests. Bacteria antagonistic to *Erwinia* are being developed as seed piece treatments for reducing seed piece decay and blackleg.

Resistant Cultivars

Plant breeding is one of the most powerful tools available for both the management of pests and the production of the best crop. Pest management is one of many factors that must be taken into account when choosing cultivars; characteristics of cultivars grown in the West are listed in Table 5.

Cultivars tolerant or resistant to disease can provide long-term, economical protection from conditions that otherwise could inflict severe losses every season. A plant is considered *resistant* to a disease if the disease organism cannot infect the plant, or if, on infection, the organism does not cause any disease symptom or measurable effect on yield or quality. The term *tolerant* is sometimes used to indicate a degree of resistance that allows the plant to tolerate a certain amount of disease with no more than

Table 5. Characteristics of Potato Cultivars in the Western States.

Cultivar	Maturity	Dormancy[a]	Tuber	Principal Uses	Possible Problems
Atlantic	Medium	3	Round, scaly skin	Chipping	Internal brown spot, hollow heart, sensitive to metribuzin
Butte	Late	3	Long, smooth, russet, shallow eyes	Fresh market	Short dormancy, sugar accumulation at low temperature, undersized tubers
Centennial Russet	Medium to Late	3	Oblong, heavy russet, shallow eyes	Fresh market	Scab, sensitive to ozone and metribuzin, sugar accumulation at low temperature
Chieftain	Medium	3–4	Oblong, smooth, red skin, shallow eyes	Fresh market	Blackleg, skinning
Kennebec	Medium to late	3	Oval, white skin, medium to shallow eyes	Chipping, home garden	Prone to higher storage losses if bruised, shallow tubers
Lemhi Russet	Medium to late	3	Oblong, russet, shallow eyes	Processing, fresh market	Blackspot bruise, hollow heart
Monona	Medium	3	Round, white skin, deep apical eye	Chipping	Seed piece decay, bacterial soft rot, blackleg, poor stands, low solids
Nooksack	Very late	6	Oblong, russet, few and shallow eyes	Fresh market, processing	Long dormancy, low yield, blind seedpieces, slow emergence
Norchip	Medium	3	Round, white skin, shallow eyes	Chipping	Prone to higher storage losses if bruised
Norgold Russet	Early	2	Oblong, russet, shallow eyes	Fresh market, home garden	Sugar accumulation, low solids, early dying, blackleg, hollow heart
Norland	Very early	2	Round, red skin, shallow eyes	Fresh market, home garden	Low solids
Red LaSoda	Early to medium	3	Round, red skin, deep eyes	Fresh market, home garden; heat tolerant	Internal necrosis, hollow heart, deep eyes
Red Pontiac	Medium to late	3	Oblong, red skin, medium eyes	Fresh market, home garden	Extremely large tubers
Russet Burbank	Late	4	Long, russet, shallow eyes	Fresh market, processing, stores well	Rough tubers, dark ended, leafroll net necrosis
Sangre	Medium	3–4	Oblong, shallow eyes, very red skin	Fresh market, stores well	Slow emergence
Targhee	Late	3–4	Oblong, smooth, heavy russet skin, shallow eyes	Fresh market, processing	Blackspot, sugar accumulation at lower temperatures
White Rose	Early to medium	2–3	Long, white skin, medium eyes	Fresh market; heat tolerant	Scab, thin skin, short shelf life, greening

a. Months required at 40°F (4.4°C) to break dormancy of seed tubers grown under normal conditions.

moderate losses. For example, the cultivar Russet Burbank may be considered tolerant to Verticillium wilt—unless the disease is severe, infected plants produce satisfactory yields as long as optimum growing conditions are provided.

Part of every breeding program is the search for resistance to serious diseases, disorders, and nematode pests. Resistance to insect pests is being investigated. New potato breeding selections are assessed for resistance to several viruses, leafroll net necrosis, root-knot nematodes, Verticillium wilt, scab, blackleg, early blight, and several physiological disorders.

The potato cultivars commonly grown in the western United States differ in their relative resistance to important diseases and disorders (Table 6). For example, Norgold Russet is highly resistant to net necrosis, while Russet Burbank is highly susceptible; however, Norgold Russet is more susceptible to blackleg and Verticillium wilt. Potato selections with high resistance to physiological disorders, net necrosis, Verticillium wilt, or root-knot nematodes are now available. Plant breeders hope to replace Russet Burbank with an equally useful cultivar that is highly resistant to most important diseases and disorders.

Cultural Practices

Proper management of the potato crop, from field preparation and planting through harvesting and storage, is essential for maximum yields of high-quality tubers. Many cultural practices, including seed selection and handling, planting, irrigation, fertilization, vine killing, and careful harvesting methods have a significant impact on pest damage. Even when you cannot choose cultural methods solely for their effect on pest management, it is important to understand their impact on pests so that you will know what to expect.

Manage your potato crop so that tuber growth begins soon enough for adequate yield, and remains uniform during the growing season. Careful water management helps prevent Rhizoctonia and seed piece decay early in the growth of the plant, reduces tuber malformations and symptoms of Verticillium wilt during the season, reduces the severity of scab, and helps prevent tuber rots as plants mature and die. Excess fertilization of indeterminate cultivars delays tuber growth and may reduce yields; inadequate fertilization also reduces yields and aggravates some diseases such as early blight and Verticillium wilt.

Sanitation. Some pest infestations begin when contaminated seed tubers, soil, water, or machinery is brought into a field. Take the following precautions to prevent the introduction or spread of potential problems.

- Use highest-quality certified seed tubers. Root-knot nematodes and damaging viral, bacterial, and fungal pathogens may be carried in contaminated tubers. Certification does not guarantee complete freedom from seedborne pests, but their spread is greatly reduced by using highest-quality certified seed.
- Thoroughly clean seed-cutting equipment with steam or pressurized water and then apply a suitable disinfectant. Table 7 lists recommended disinfectants; be sure to keep surfaces moist with disinfectant for the recommended time and to rinse off corrosive disinfectants. Disinfect at least once a day and *always between seed lots.*
- Do not irrigate with tailwater from fields infested with root-knot nematodes or other soilborne pathogens or pests; do not use water that may carry runoff containing harmful herbicides.
- If surface water is used for irrigation, consider putting in settling ponds between the canal or ditch and the field to reduce the spread of nematodes.
- If irrigation water comes from canals or ditches, consider installing screens to filter out seeds, rhizomes, and other weed parts.
- Clean equipment between fields to avoid moving contaminated soil.
- Schedule discing operations to maximize the destruction of culls, harmful pests that survive or reproduce in tubers, and insects that pupate in the soil. In areas with cold winters, discing after the first killing frost usually destroys the most culls and leaves the fewest potato volunteers. Disc immediately after harvest in warmer areas, being sure culls are buried to reduce tubermoth activity where this pest occurs. Rototilling is sometimes used when large numbers of culls are expected to survive.
- Make sure residue from previous crops has decayed before planting. Several pathogens survive between seasons on crop debris.
- Use roguing or spot treatments with herbicides to destroy problem weeds before they produce seed, especially dodder, nightshades, and perennials that are hard to control.
- Destroy weed stands along field borders; if allowed to mature, they produce seeds that are a source of infestations in the cultivated area. A number of weeds can be reservoirs for insect pests or pathogens of potatoes.

Crop Rotation. Proper crop rotations enhance soil fertility, help maintain soil structure, reduce certain pest problems, increase soil organic matter, and conserve soil moisture. Herbicides not available for potatoes can be used in certain rotation crops to control problem weeds. Whenever possible, use rotations that reduce problem pests and avoid rotations that may increase them.

Table 6. Relative Resistance of Potato Varieties Grown in the Western States to Common Diseases and Disorders.

| VARIETY | Common Scab | Net Necrosis | Early Blight | | Verticillium Wilt | Blackleg | Virus X | Growth Cracks | Hollow Heart | Second Growth | Storage and Seed Piece Decay | Other[c] |
			Foliage	Tuber[b]								
Atlantic	S	R	S	MS	MR	MS	HR	MR	S	R	HS	S to ibs
Butte	R	R	MS	MR	MS	S	HR	MR	R	R	R	MR to bb
Centennial Russet	S	R	MS	MR	MS	S	S	S	S	R	MR	S to o
Chieftain	MR	R	MR	MR	MS	S	S	MR	R	R	MS	–
Kennebec	S	R	MR	MS	MS	S	S	S	S	R	S	MS to bb
Lemhi Russet	R	MR	MR	MR	MR	MR	–	MR	S	MR	MS	S to bb
Monona	MR	R	S	MS	MS	S	R[d]	MR	MR	R	S	MS to bb
Nooksack	HR	R	MR	MR	MR	MR	S	MS	MS	R	S	HR to bb
Norchip	MS	R	HS	HS	MS	S	S	MS	R	R	MS	MR to bb
Norgold Russet	R	R	S	R	S	S	S	MR	S	R	S	MR to bb
Norland	MS	R	MS	MR	S	S	S	MR	R	R	MS	MR to bb
Red Pontiac	S	R	MS	MR	MR	S	S	MS	S	R	MS	–
Red LaSoda	S	R	MR	MR	MS	MS	S	MR	S	MR	MR	S to ibs
Russet Burbank	R	HS	MS	MR	MS	MR	S	MS	MS	HS	MS	S to bb
Sangre	MS	R	S	MR	S	MS	MR	MR	MR	R	MR	MR to bb
Targhee	HR	S	MS	MR	MR	MR	HR	S	MS	R	S	S to bb
White Rose	MS	R	MS	MS	MR	MS	MS	S	MR	MR	MS	MR to bb

The header spans: SUSCEPTIBILITY TO DISEASE OR PHYSIOLOGICAL DISORDER[a]

a. R = Resistant, S = Susceptible; H = Highly, M = Moderately
b. Immature tubers are more susceptible.
c. ibs = internal brown spot, bb = blackspot bruise, o = ozone
d. Resistant to mild mosaic and rugose mosaic.

Generally, the most useful rotations for potato fields are forage crops and grains, including corn. Rotation crops that are useful only in some areas include alfalfa, cotton, sugar beets, beans, and Sudangrass. Grain and hay crops reduce Pacific Coast wireworm and northern root-knot nematode, but they increase wheat wireworm and Columbia root-knot nematode. Alfalfa rotations reduce wireworms and Columbia root-knot nematode, but increase northern root-knot nematode. The benefit of a rotation crop may be increased by selecting cultivars of that crop that are more resistant to problem pests.

Seed Tuber Handling. The longer a seed tuber is stored or aged after its dormant period is over, the more sprouts it produces, and consequently the more mainstems. Aging usually occurs during the time seed tubers are stored, generally by the seed producer. Some aging is desirable to break the dormancy of enough eyes so that each seed piece will produce two or three stems. Cultivars such as Nooksack, which has a long dormancy period, must be aged longer than Russet Burbank to ensure a sufficient number of sprouts. However, if seed tubers are stored too long they will give rise to less vigorous plants that produce smaller tubers and lower yields.

Store seed tubers at 35° to 38°F (1.7° to 3.3°C). Approximately 2 weeks before cutting, warm seed tubers gradually to 50° to 55°F (10° to 13°C) and hold them at

Table 7. Disinfectants Commonly Recommended for Potato Handling Equipment and Storage Facilities.[a]

Disinfectant	Effectiveness for		Inactivation by		Corrosiveness	Safety	Concentration	Exposure Time	Shelf Life	Comments
	Wet Bacteria and Slime	Dry Bacteria and Slime	Organic Matter	Hard Water						
Quarternary Ammonium Compounds	Excellent	Excellent	Only slightly	No	Slight	Use caution.	Label directions	10 min	1–2yr	Diluted disinfectant relatively safe, concentrated form is poisonous. Stainless.
Hypochlorites (5.25% bleach)	Excellent	Excellent	Yes	No (except iron)	Yes	Irritant, caustic.	1:50(0.1%) or 1:200	10 min	3–4 mo. undiluted	Quick acting, inexpensive; caustic to skin and clothing. Use at 1:50 when mixing with water only. For maximum effectiveness use 1 part 5.25% bleach: 200 parts water: 0.6 parts white vinegar.
Iodine Compounds	Excellent	Excellent	Some; not greatly	No (except iron)	Yes	Relatively safe. Use caution.	Label directions	10 min	1–2 yr	Do not take internally. No longer effective if it loses yellow-brown color. Tamed iodophor compounds work best.
Phenolic Compounds	Excellent	Excellent	Some; not greatly	No	No	Oral poison. Use caution.	Label directions	10 min	1–2 yr	Provides residual action. These have name "phenol" on label of ingredients.
Formaldehyde	Good	Poor	No	Yes	No	Unsafe vapors. Use caution.	0.37–1.0%	30 min	1–2 yr	Use may be cancelled. Gives off irritating, choking fumes. Not generally recommended.
Copper Sulfate	Good	Good	No	Yes	Yes	Use caution.	10lb/100 gal. water	30–60 min	>10 yr. as solid	Not widely used; used mostly as a soak for crates and bags.

a. Adapted from: Disease Control Guidelines for Seed Potato Selection, Handling, and Planting, Extension publication PP-877, North Dakota State Univ.

Registrations may vary; check with local authorities.

that temperature with a relative humidity greater than 90% and good ventilation. This encourages wound healing (suberization) when they are cut, greatly reducing the incidence of seed piece decay after planting. If tubers are cut when they are just beginning to sprout, a stage sometimes called *peep* or *peek*, emergence is more rapid and you can more easily choose a seed piece size that gives the number of sprouts you want. Cut seed tubers before sprouts exceed about 1/8 inch in length to avoid breaking them. If sprouts are broken, spread of mechanically transmitted viruses is more likely and seed pieces may develop multiple sprouts, which are weaker and may form too many stems per hill.

A seed piece size of 1 1/2 to 2 1/2 ounces is recommended for optimum performance in most areas. The larger size is recommended for cultivars that have few eyes, such as Centennial, Kennebec, Nooksack, and Norgold. Cut seed in an area out of drafts, and wet down the area to increase the humidity. Follow good sanitation practices during cutting; clean and disinfect cutting equipment thoroughly between seed lots (Table 7). Protect the cut seed from sun and wind when hauling.

Plant cut seed pieces immediately in moist soil (60% to 80% of field capacity) that is at a minimum of 45°F (7°C) to accelerate emergence and wound healing after planting. If you cannot plant cut seed immediately, hold it at 50° to 55°F (10° to 13°C) with good aeration and high humidity to speed wound healing. Do not store cut seed in bulk trucks.

Seed Treatment. Seed piece decay frequently involves a *Fusarium* fungus acting synergistically with bacteria. Therefore, chemical seed treatments, which primarily act as fungicides, are useful when conditions favor development of *Fusarium* on seed pieces. Treatments that control bacteria involved in seed piece decay are being developed, but none are commercially available at the present time.

When good quality seed tubers are handled carefully and field conditions remain favorable for rapid wound healing and emergence, chemical seed treatments are generally not necessary. Depending on growing area and planting date, some growers can safely avoid the expense of seed treatments, while others may benefit from the insurance provided by applying them regularly. To decide if seed treatments may be of benefit, you must understand the conditions that promote seed piece decay and you must know about their frequency and severity in your situation. Important points to consider include:

- How confident can you be that soil moisture at planting will be optimal?
- Are the cultivars you plant resistant or susceptible to seed piece decay?
- How well controlled is your seed-cutting operation? Can you avoid conditions unfavorable to wound healing before planting?
- How predictable are the condition and quality of the seed tubers you use?
- What is your planting time? What is the likelihood of unfavorable postplanting weather conditions?

Planting. Proper field preparation is essential for maximum quality and yield. Residues from previous crops must be thoroughly worked up and incorporated to allow complete decomposition. Excessive undecomposed organic matter favors scab development. Work finer-textured soils in the fall where possible; work sloping land as late as possible in fall or early in spring to avoid erosion. A winter cover crop is necessary in some areas to prevent erosion of coarse soils by wind; work these fields in the spring. If spring wind erosion is a problem, leave the crop residue on or near the soil surface. Use alfalfa plowshares when working up alfalfa to be sure old crowns are killed; shoots from old alfalfa plants are likely to be reservoirs of alfalfa mosaic virus.

Plow depth and equipment vary with location and conditions. Plow deep soils to a depth of 8 to 12 inches (20 to 30 cm); plow shallow soils no deeper than 1 inch below the plow sole. A plow width of 14 inches (36 cm) or more is satisfactory for potato fields. If a lot of residue must be turned under, use a plow with a 16- to 18-inch bottom and high clearance. Chiseling is recommended in many areas to break up compacted soil and ensure proper drainage. Level the field so water can be applied uniformly without ponding. Basin tillage, which leaves dams spaced periodically within furrows, may be desirable on uneven land to make sure water penetrates evenly. Beds prepared for potato planting are usually flat, but may also be raised. Preplant fertilizer may be mixed into the soil at the time of bed preparation.

Plant seed pieces in soil that has a moisture level of 60% to 80% of field capacity from the level of seed pieces to 2 feet (61 cm) below the surface. Use a preplant irrigation if soil is dry. This will result in a much higher and healthier plant stand than if irrigations are applied between planting and emergence. Be sure the soil temperature is at least 45°F (7°C) before planting. A soil temperature of 50° to 55°F (10° to 13°C) at planting depth is ideal for encouragement of vigorous sprout emergence, but in many areas waiting for the temperature to reach 50°F may mean losing valuable time. Planting in cold soil or soil that is either too wet or too dry increases the likelihood of damage by Rhizoctonia, blackleg, and seed piece decay.

Planting depths vary from 4 to 6 inches (10 to 15 cm), depending on the area, time of year, and conditions at planting. Seed pieces are planted deeper in coarse soils than in finer-textured soils. Deeper plantings are sometimes used for frost protection. Generally, shallower planting is used where emergence is expected to be slower. Be sure hilling operations give a final seed piece depth of 6 to 8 inches (15 to 20 cm) by the time tubers begin to form.

Plant seed pieces in rows that are 30 to 36 inches (75 to 90 cm) wide; the width used is usually determined by local cropping patterns and equipment. Spacing of seed pieces within rows may be used to control the number of stems per hill and tuber size. Potatoes grown for seed are often planted closer together to produce larger numbers of smaller tubers. In commercial production, spacing is influenced by cultivar. Russet Burbank is usually planted at spacings of 9 to 12 inches (23 to 30 cm). Spacings of 5 to 6 inches (13 to 15 cm) are used for Kennebec, because it tends to produce fewer tubers per hill, and narrower spacing helps keep tubers from growing too large. Red cultivars are planted 6 to 8 inches (15 to 20 cm) apart for the same reason. Norgold Russet spacings are usually 8 inches (20 cm), White Rose 7 to 8 inches (18 to 20 cm), and Centennial Russet 8 to 9 inches (20 to 23 cm). Cultivars such as Norchip that set large numbers of tubers are spaced farther apart, from 11 to 14 inches (28 to 36 cm). The quantities of seed tubers needed for different row widths and seed spacings are listed in Table 8.

Fertilizer applications are made at planting if preplant applications are not used or only a portion is applied preplant. Fertilizer bands are placed at least 2 inches (5 cm) to the sides and 2 inches below the seed pieces. Otherwise, seed pieces may be damaged by high salt concentrations, causing stand reductions. Pick and cup planters are the most popular types used for potatoes (Figures 12 and 13). Keep the planter in good operating condition and check to be sure it is adjusted to work properly with the seed size you are planting. With a cup planter, use cups that are the correct size; with a pick planter, use pick lengths and an arrangement suited to the seed size.

When plants begin to emerge, count the number in 100-foot sections and compare to the expected number for the spacing you used (Table 8). Where the number of emerged plants is significantly lower than expected, dig up seed pieces to determine the cause of the reduced stand. Poor stands may be caused by mechanical errors in planting, blank seed pieces (lacking eyes), exposure of seed to sprout inhibitors, disease, or damage during cultivation.

Table 8. Number of Seed Pieces Used per 100 Feet of Row, and Total Required per Acre for Different Planting Dimensions (seed piece size 1.5 ounces).

Seed Piece Spacing Inches (cm)	Number per 100 feet (30.5 m)	Total (Cwt) Required Per Acre for Row Widths: Inches (cm)			
		30 (76)	32 (81)	34 (86)	36 (91)
6 (15)	200	32.6	30.7	28.8	27.1
8 (20)	150	24.5	22.9	21.6	20.4
10 (25)	120	19.6	18.4	17.3	16.3
12 (30)	100	16.3	15.4	14.4	13.5
14 (36)	86	14.0	13.1	12.4	11.6

Figure 12. A picker planter uses metal spikes or picks, available in various lengths, to transfer seed pieces from the bin into the seed bed. Picks can spread pathogens from infected to uninfected seed pieces.

Hilling. Perform hilling operations after plants begin to emerge and do not cover the emerging plants. Discs, rolling cultivators, hilling listers, or implements with winged cultivator teeth are used. Hilling allows the use of a shallow seed piece depth for rapid emergence, while providing the soil depth necessary later in the season for proper tuber development and reduced exposure to sunlight and adverse temperatures. Exposed tubers develop green color due to chlorophyll production and bitter-tasting compounds called glycoalkaloids in the skin and cortex; they are also susceptible to tuberworm infestation. Shallow tubers are more susceptible to heat necrosis. For best protection of tubers, be sure hills are flat and broad rather than narrow and peaked. Cultivation during hilling destroys emerged weeds, and some soil-applied herbicides can be mixed in at this time. When applying herbicides during hilling operations, be sure not to cover sprouts with treated soil; set equipment to throw treated soil back into the furrow. Plan to complete hilling before plants are so large that pruning of stolons or roots may occur.

Drag-off is the practice of removing the tops of hills to hasten emergence, usually when seed pieces have been planted in raised beds. It is not widely used and is not recommended because of the potential for damaging sprouts.

Frost Protection. In growing areas where frost may occur after plants are up, irrigation can be used to keep the foliage from freezing. Solid set sprinklers are usually required; they must be turned on before the air temperature drops below freezing, and left on until the air temperature has risen and the threat of freezing has passed. As water freezes, it releases a small amount of heat, which is sufficient to keep the leaf tissue from freezing. However, new drops of water must freeze continuously throughout the frost period to keep foliage above the freezing point.

Irrigation. Availability of soil water is a major factor that determines yield and quality of the potato crop. Too little water will reduce yields, induce tuber malformations, or increase severity of scab or Verticillium wilt symptoms. Excess or poorly timed irrigation may reduce yields and quality, cause several disease problems in the field or in storage, or leach nutrients from the root zone. Fluctuations in water availability favor disorders such as second growth and internal necrosis.

Efficient irrigation requires finding out how much *available water* the soil can hold. Available water is that portion of the soil water that can be withdrawn by plants (Figure 14). During the growing season, irrigation is needed when a certain proportion of the available water, the *allowable depletion*, has been used. The allowable depletion in a particular field varies according to soil type, stage of crop growth, total amount of available water, weather conditions, and irrigation cost. Because potatoes are sensitive to water stress, the allowable depletion is no more than 30% to 40%. To minimize scab infection, the allowable depletion is no more than 20% during tuber initiation.

Potatoes are a shallow-rooted crop; 90% of the roots grow in the top 12 to 18 inches (30 to 45 cm) of the soil. You can determine from the clay content and soil texture in the top 18 inches how much available water the soil can hold (Figure 15). For a soil profile that includes layers of different soil types, calculate the available water separately for each layer; then add them together to obtain the total available water in the rooting zone (Figure 16). Take salinity into account in estimating available water; if the soil or irrigation water contains high levels of salts, plants will be able to withdraw less water, so a smaller proportion will be available to plants than that estimated using Figure 16.

Irrigation Methods. Most potatoes in the western states are irrigated with sprinklers. Center pivot, wheel line, and solid set systems are most commonly used. Sprinkler systems provide the most flexibility and most efficient water application, and fertilizers and some pesticides can be applied through sprinklers. Sprinklers are readily adapted to uneven ground. When preparing fields, be sure not to leave any low spots where water will collect. Sprinkler irrigation provides conditions in the potato canopy that are favorable for certain diseases, such as early blight, late blight, and white mold (Sclerotinia). To reduce spread of these diseases, allow foliage to dry out between irrigations.

Furrow irrigation may be used where the slope of a field is less than 2% and where rows are not longer than 600 to 800 feet (182 to 244 m). The length of furrows is affected by soil texture. Uniform application of water and fertilizer is more difficult with furrow systems. Tuberworm damage is almost always higher where furrow irrigation is used within the range of this pest, because soil cracking is more prevalent. Powdery mildew development is favored by furrow irrigation, translucent end occurs more frequently, and the potential for Verticillium wilt is higher than with sprinkler irrigation. Early blight is less favored by furrow irrigation.

Subirrigation can be used where the water table can be raised easily, soil is of uniform texture, and fields are relatively level. The low, flat, peat soils of the Sacramento-San Joaquin Delta area of California are favorable for the use of subirrigation.

Figure 13. A cup planter uses cups, available in different sizes, to transfer seed pieces into the seed bed. Both picker and cup planters can be used to apply fertilizer during planting.

Soil is about half solid material by volume (large circle). The rest of the soil volume consists of pore spaces between soil particles. Pore spaces hold varying proportions of air and water (small circle).

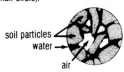

When the soil is **saturated** after irrigation or rain, pore spaces are filled with water.

When soil has drained following irrigation, it is at **field capacity**. In most soils, about half of the pore space is filled with water. About half of this water is **available** to plants; the rest is unavailable because too much suction is needed to remove it from pore spaces.

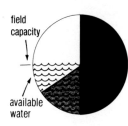

The **allowable depletion** is the proportion of the available water the crop can use before irrigation is needed.

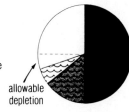

At the **wilting point**, all available water is gone. Plants die unless water is added.

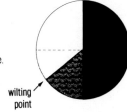

Figure 14. The soil reservoir.

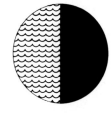

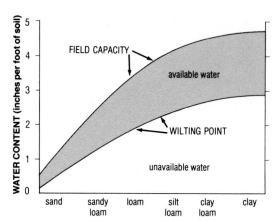

Figure 15. The higher the percentage of clay in soil, the greater the water-holding capacity. The shaded area represents available water.

Preirrigation. Soil moisture should be at 60% to 80% of field capacity at planting time. If rainfall is not adequate to fill the soil reservoir, then use a fall irrigation, or irrigate before planting. Avoid irrigations between planting and emergence. Irrigations at this time can increase blackleg, Rhizoctonia, and seed piece decay. It is best to have soil moisture high enough so that the first irrigation is not needed until plants have emerged. However, irrigations must be used if the soil becomes excessively dry.

Postplant Irrigations. With sprinklers, each postplant irrigation should bring the top 18 inches (45 cm) of soil back to field capacity; do not irrigate to a depth of more than 24 inches (60 cm). The timing and amounts of postplant irrigations depend on the water-holding capacity of the upper 18 inches of soil and the rate at which the water is used by evapotranspiration. Maintaining adequate soil moisture is critical during the tuber initiation and tuber growth phases; water stress during these periods may cause tuber malformations and translucent end, especially in Russet Burbank. Dry conditions during early tuber development favor scab infections.

Final irrigations in most growing areas should be timed to allow soil moisture to drop to about 60% of field capacity at the time of vine killing. This level of soil moisture encourages proper development of tuber skin and decreases the chance that tubers will be infected during harvest by early blight, late blight, or soft rot pathogens. However, if soil moisture falls below 50% during vine killing, stem-end browning may result. In hot growing areas, light irrigations may be continued until harvest to keep soil temperatures down. Excess irrigation at this time may reduce oxygen levels in the soil and cause tuber rot or black heart.

Irrigation Scheduling. Timing of irrigations during the growing season can be based on various measures of soil moisture together with a water budget, which can be based on evapotranspiration or pan evaporation data. A combination of methods is usually best.

Always check soil moisture before applying water and estimate how much available water remains in the crop rooting depth. Use a soil tube or shovel to take soil from the rooting zone at several points in each field. In furrow-irrigated fields, check a few places at the head end of the runs, some in the middle, and some at the tail end. Use Table 9 as a guide for judging the depletion level in soil taken from the root zone.

Water needs usually vary from one part of a field to another, especially if the field includes different soil types or slopes. Plants in a sandy streak or where root development has been restricted will show stress sooner than the rest of the crop. Watch these weak areas to gain advance notice of when irrigation is needed for the rest of the field. However, schedule irrigations according to the need shown by most of the crop.

Instruments are available that measure the moisture content of the soil. Tensiometers and neutron probes are frequently used for monitoring soil moisture. To obtain reliable readings, you must install these instruments in areas representative of the field, including spots where water stress occurs more readily. Aerial infrared photography can help identify areas of different moisture stress within fields. At each site install one soil moisture probe at the rooting depth of the current growth stage and a second probe 18 to 24 inches (45 to 60 cm) deep. Follow the recommendations of your supplier and your extension agent, farm advisor, or other irrigation experts in using soil probes for irrigation scheduling.

A convenient way to monitor soil moisture indirectly is to use a *water budget* (Figure 17) to estimate how much water the crop uses from day to day under prevailing weather conditions. After soil has drained to field capacity, further loss of soil water occurs mainly through

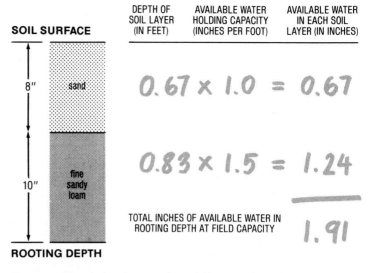

Figure 16. To calculate how much available water the soil can hold at field capacity, first examine the soil profile to see if there are layers of different soil textures in the rooting depth of the crop. For each soil layer, multiply the depth of the soil layer in feet by the available water-holding capacity of that soil texture (*see below*). Add the results from all the layers to get the total available water-holding capacity for the soil profile.

Sand	1.0 inches/foot
Sandy Loam	1.5 inches/foot
Loam, Silt Loam, Clay	2.2 inches/foot

Table 9. Judging Depletion of Soil Water by Feel and Appearance. Numbers in each column are inches of water needed to restore 1 foot of soil depth to field capacity when soil is in the condition indicated.

Coarse-Textured Soils	Inches of Water Needed	Medium-Textured Soils	Inches of Water Needed	Fine-Textured Soils	Inches of Water Needed
Soil looks and feels moist, forms a cast or ball, and stains hand.	0.0	Soil dark, feels smooth, and will form a ball; when squeezed, it ribbons out between fingers and leaves wet outline on hand.	0.0	Soil dark, may feel sticky, stains hand; ribbons easily when squeezed and forms a good ball.	0.0
Soil dark, stains hand slightly; forms a weak ball when squeezed.	0.3	Soil dark, feels slick, stains hand; works easily and forms ball or cast.	0.5	Soil dark, feels slick, stains hand; ribbons easily and forms a good ball.	0.7
Soil forms a fragile cast when squeezed.	0.6	Soil crumbly but may form a weak cast when squeezed.	1.0	Soil crumbly but pliable; forms cast or ball, will ribbon; stains hand slightly.	1.4
Soil dry, loose, crumbly.	1.0	Soil crumbly, powdery; barely keeps shape when squeezed.	1.5	Soil hard, firm, cracked; too stiff to work or ribbon.	2.0

evaporation from the soil surface or transpiration from leaves. The combination of evaporation and transpiration is called *evapotranspiration* (ET). If you know how much available water is in the crop rooting depth at field capacity and how much water is lost through ET each day, you can estimate the amount of available water remaining at any time by adding up the daily ET values. Newspapers and radio and television stations in some areas report ET figures for major crops, including potatoes. Computer programs that provide ET data and help calculate water budgets are available in some areas.

You can also use daily *pan evaporation* and a *water use curve* for the crop to develop your own ET figures. Figure 18 shows a water use curve for potatoes, representing water use at different stages of growth. To use this curve, determine the water use coefficient for the current stage of potato growth, for example a value of 0.45 during tuber initiation. Multiply the daily pan evaporation by this figure to get actual crop water use. Pan evaporation can be obtained from local weather stations, or you can set up a measuring device in or near your own fields. A standard U. S. Weather Bureau Class A evaporation pan can be used, or a No. 1 or No. 2 galvanized washtub set up in the same way. Set the pan 5 inches above the ground and fill it with water to 1 inch (2.5 cm) below the top; measure the amount evaporated at the same time each day and refill to the 1-inch level.

Water budgets provide estimates of crop water use. However, actual water use is affected by disease, weeds, insects, physical characteristics of individual fields, and management factors. Use a direct measurement of soil moisture to make the final decision about when to irrigate.

A good irrigation program follows these basic guidelines.

- Start with soil at field capacity, and avoid irrigation between planting and emergence.
- Monitor the soil moisture.
- Record daily ET.
- Keep records of irrigation and rainfall amounts.
- Do not irrigate deeper than 18 to 24 inches (45 to 60 cm) or more frequently than necessary.
- Avoid wet soil (soil with moisture in excess of levels recommended for the current stage of growth.)
- Use irrigation design that gives uniform water distribution.

Managing Salinity. Prevention of harmful salt buildup is an important part of irrigation. As part of routine soil analysis, have a laboratory determine the concentration of total salts and the proportion of sodium in the soil. A soil salinity reading of 2 to 5 mmho/cm indicates a possible salinity problem. If the salt concentration in the soil is high, plan to leach the salt below the root zone with extra water during preirrigation. Consult your local ex-

FIELD NO. 14

ALLOWABLE DEPLETION 40% of 2.0 in. = 0.80 inch

DATE	ET IN INCHES PER DAY	CUMULATIVE ET	
7/10			irrigation
7/11	0.23	0.23	
7/12	0.25	0.48	
7/13	0.27	0.75	check field
7/14	0.26		irrigation
7/15	0.27	0.27	
7/16	0.25	0.52	
7/17	0.24	0.76	check field
7/18	0.26		irrigation
7/19	0.28	0.28	
7/20	0.27	0.55	
7/21	0.25		irrigation

Figure 17. If you have a source of evapotranspiration (ET) data, you can use a water budget to estimate the interval between irrigations. You can also use pan evaporation and the crop water use coefficient (see Figure 18 and text) to develop your own ET figures. Keep a running total starting the day after irrigation. When the total approaches the allowable depletion, check soil moisture in the field to make a final decision on when to irrigate.

tension agent, farm advisor, or other experts about guidelines for your situation.

Test irrigation water for total salts and sodium every 2 or 3 years, especially if it comes from wells where the water table is dropping. Test each well separately, as salt concentrations may differ greatly, even among wells in the same area.

Fertilization. Adequate nutrient availability throughout the growing season is necessary for the best yield and quality. If nutrient deficiencies occur during tuber growth, the plant shunts nutrients from the stems and leaves to the growing tubers, thereby hastening aging of the vines. As a result, certain diseases such as early blight and Verticillium wilt are aggravated, and yields are reduced. On the other hand, excess fertilizer delays the onset of tuber growth in indeterminate cultivars and may reduce their yields; tuber decay after harvest may also be increased, and processing qualities such as specific gravity may be lowered.

Potato plants use several nutrients from the soil during growth. Amounts removed are directly related to yield (Table 10); that is, a yield of 600 hundredweight per acre removes about twice as much of the soil nutrients as 300 hundredweight per acre. Nutrients used in the smallest quantities (zinc, manganese, iron, copper, and boron) are referred to as micronutrients. Some nutrients are used in large enough quantities or are so limited in their availability that they need to be added to grow a successful potato crop.

Nitrogen almost always is needed; phosphorus and potassium usually are needed. Calcium and magnesium are needed on some acid soils. Zinc and manganese may be required on some calcareous, alkaline soils in the Northwest; they are applied as chelated micronutrients so they will not be made unavailable to the plant by chemical reactions in alkaline soil. Sulfur supplements are required in areas where the soil or water is low in sulfur. Muck or peat soils may benefit from additions of copper and manganese.

To determine fertilizer needs, you must consider cropping history and individual field history, records of petiole analysis from previous seasons, preplant soil analysis, and petiole analysis during the season. For nutrients that can be added during potato growth (e.g. nitrogen and zinc), petiole analysis is very useful for determining when additions are needed. Petiole analysis also provides a useful guide for the next season, by indicating which nutrients are being depleted. Soil tests help determine the need for preplant applications of nitrogen, phosphorus, potassium, sulfur, and other nutrients. Sampling techniques and fertilizer recommendations are presented in several references listed at the end of this book.

Fertilizer is used most efficiently when supplied closest to the time it is needed. This is especially true of nitrogen, which can be lost readily by leaching or microbial activity. Nutrients are generally needed to a greater extent during early and middle growth, to assure maximum vine and root growth and tuber initiation and early growth. Late application of nitrogen delays tuber growth and maturity. Because growing conditions and cultivars differ from one area to the next, guidelines for adequate nutrient ranges differ. Consult the References for specific information and be sure to check with your extension agent, farm advisor, or other experts to get the latest guidelines for your area.

Most nutrients can be applied before planting in sufficient quantities to last all season. Phosphorus, potas-

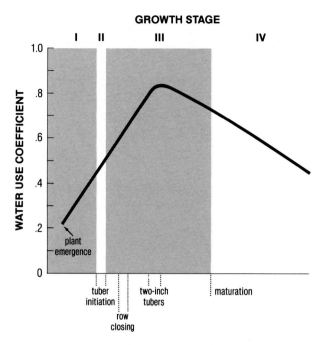

GROWTH STAGE

Figure 18. Water use curve for potatoes (adapted from *Crop Water Use Curves for Irrigation Scheduling*, Special Report 706, Agricultural Experiment Station, Oregon State University, Corvallis, OR). Daily water use (ET) can be calculated by multiplying the daily pan evaporation (available from newspapers or weather stations) by the water use coefficient for the current crop stage as indicated in this table. These ET figures can be used in a water budget such as the one shown in Figure 17.

Table 10. Mineral Nutrients Removed from the Soil by Potato Vines and Tubers.

NUTRIENT	AMOUNT REMOVED (LB/ACRE) BY DIFFERENT YIELDS (Cwt/Acre)				
	Vines	300	400	500	600
Nitrogen	139	128	171	214	257
Phosphorus	11	17	23	29	35
Potassium	275	144	192	240	288
Calcium	43	4.4	5.9	7.4	8.9
Magnesium	25	8.9	11.8	14.7	17.6
Sodium	2.70	1.74	2.32	2.90	3.48
Zinc	0.11	0.11	0.14	0.18	0.22
Manganese	0.17	0.04	0.06	0.07	0.08
Iron	2.21	0.79	1.06	1.32	1.58
Copper	0.03	0.06	0.08	0.10	0.12
Boron	0.14	0.04	0.05	0.06	0.07

sium, and most micronutrients do not move readily with soil water; they should be applied at or before planting. Whether fertilizer is applied as a preplant broadcast or banded at planting depends on soil type, soil temperature, and fertilizer formulation and cost. Banding near the seed piece assures rapid availability and uptake under adverse conditions; broadcasting distributes the nutrients more completely throughout the root zone.

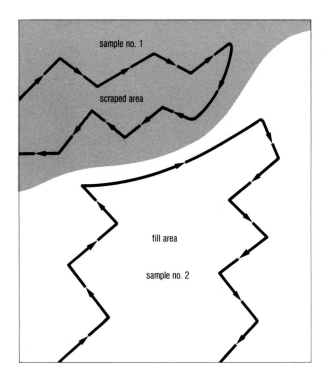

Figure 19. Suggested sampling pattern for taking petioles, to be used in nutrient analysis, from a nonuniform field.

Because nitrogen is easily leached from the rooting zone, especially in coarse-textured soils, it is best to apply only a portion of the total requirement before planting. The amount applied before planting depends on cultivar, soil type, and growing conditions. If leaching is not a problem, most or all of the seasonal requirement can be applied before or during the planting of determinate cultivars such as Norgold Russet. If you are equipped to apply fertilizer through irrigation water, apply no more than half the total nitrogen requirement to Russet Burbank or other indeterminate cultivars; otherwise, tuber growth will be delayed and yields may be reduced. Use petiole analysis to schedule application of the remainder of the seasonal requirement.

Soil Testing. A preplant soil test is the best way to determine what nutrients should be applied before planting. Consult the References about soil sampling, and follow the recommendations of your local extension agent, farm advisor, or other experts and the testing laboratory for taking soil samples. The best time to sample is between crops. Be sure to take samples that represent any differences in soil conditions or cropping history within a field. Soil tests can also indicate if there is a need for calcium, manganese, sulfur, or any other micronutrient.

Soil test results for nitrogen are not as accurate as for the other nutrients because the amount of nitrogen available to plants can change rapidly. Some areas do not recommend basing seasonal nitrogen applications on soil test results. For specific fertilizer recommendations based on soil test results, consult the References concerning your area.

Petiole Analysis. The crop nutrient status during the season can be assessed by measuring nutrient concentrations in a portion of the vines. The most reliable way to do this is to take adequate and representative samples of petioles and have a reliable laboratory in your area analyze them for nutrient levels. Take the first sample when tuber initiation begins; in Russet Burbank this is about the time plants are at the 8-leaf stage and about 10 inches (25 cm) tall. Follow a pattern like the one illustrated in Figure 19, removing one petiole from each plant sampled. Always take the first fully expanded leaf below the top, usually the fourth leaf (Figure 20). Remove the leaflets from each petiole and store all the petioles for one sample in a clearly marked paper bag. Good results are obtained with 40 petioles per sample and at least three samples per field. Deliver samples to the laboratory immediately; if you cannot, dry them at a temperature of 100° to 150°F (38° to 66°C) or store them in a freezer until you can. Take at least three or four samples between tuber initiation and early maturation. Some areas recommend sampling at least once every 7 to 14 days. Keep records of all the results; they will be helpful in planning fertilizer applications for next season. Recommendations for adequate levels of major nutrients (nitrogen, phosphorous, potassium) based on petiole analysis are available for most areas, and recommendations for micronutrient levels are also available for some areas. Recommendations based on whole-leaf analysis are available in some areas. Check with local authorities and consult the publications for your area listed in the References.

Use petiole test results to schedule midseason nitrogen applications. Petiole nitrogen concentrations in Russet Burbank should be kept at the level considered adequate for midseason until about 3 weeks before vinekill. If nitrogen is not allowed to decrease at this time, tubers will not mature properly; bruising and decay losses may be increased, specific gravity may be low, and sugar content may be high.

Guidelines for adequate midseason petiole nitrogen vary with area, application method, and cultivar. In some cases the market destination of the crop (i.e., chipping, french fries, fresh market, seed) affects fertilizer guidelines. Contact your local extension agent, farm advisor, or other experts to learn the recommendations for your area.

Apply nitrogen through sprinklers as urea, ammonium, ammonium nitrate, or combinations of these. Use caution when irrigating; overirrigation may leach nitrate from the root zone. Make the first application at the time of tuber initiation, then use the trend shown by petiole analysis to schedule subsequent applications to keep the petiole level adequate (Figure 21 shows an example using Idaho guidelines). Making the final application about 4 weeks before vinekill usually allows nitrogen to decrease properly. However, in some soils petiole nitrogen decreases more quickly, and an additional nitrogen application is needed. Continue to monitor petiole nitrogen after the final fertilizer application so that you will know if additional treatment may be necessary.

Petiole analysis may also indicate whether phosphorus levels are likely to become deficient during the season. Research in Idaho indicates that Russet Burbank yields are improved if petiole phosphorus is kept above 1,000 parts per million until 20 to 30 days before vinekill. Because petiole phosphorus decreases during the season at a uniform rate, a graph can be used to predict whether petiole phosphorus is going to drop below the critical level too soon (Figure 22). If so, applications of a highly soluble form of phosphorus, such as ammonium polyphosphate, are recommended in Idaho; other areas do not recommend midseason phosphorus applications. The most widely recommended way to keep phosphorus above the critical level throughout the season is to apply adequate phosphorus before or during planting. Guidelines based on petiole phosphorus are available for areas other than Idaho. Consult the References and check with your local experts for details.

Vine Killing. In some areas, particularly in the case of early crops that go directly to fresh market or processors, potatoes are harvested while the vines are still green. Most potatoes, however, including Russet Burbank, are harvested after the vines have died. Harvesting after vine death provides several benefits. Tubers are allowed to mature and develop thicker skin (periderm) so they are less susceptible to bruising and diseases. Specific gravity and sugars reach more desirable levels, which is especially important for potatoes to be processed. Inoculum of some diseases such as late blight is reduced by allowing vines to dry up before harvest. In the case of seed crops, tuber size can be controlled and virus-transmitting aphid populations can be avoided by killing vines early.

Vine death occurs naturally in areas with early frosts and in hot areas where cutting back on water will rapidly kill vines. In most areas, mechanical or chemical methods of vinekill are used. Vinekill may cause vascular discoloration in tubers. It is usually confined to the stem end but sometimes spreads throughout the tuber. Vascular discoloration is more likely to occur if vine-killing agents are applied when soil moisture is below 50% of field capacity.

Follow these guidelines to reduce the risk of injury and increase the effectiveness of vine-killing treatments.

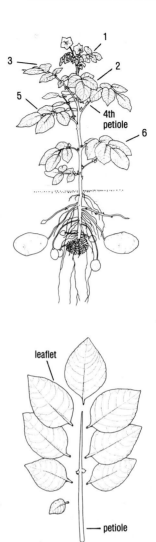

Figure 20. When taking petiole samples for nutrient analysis, select the fourth petiole (the first fully expanded leaf) from the growing tip (*top*). Remove all leaflets from the petiole (*bottom*) before sending samples to be analyzed.

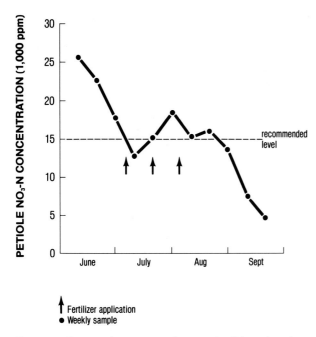

Figure 21. Pattern of nitrogen applications (40 lb/acre) based on petiole samples in southcentral Idaho.

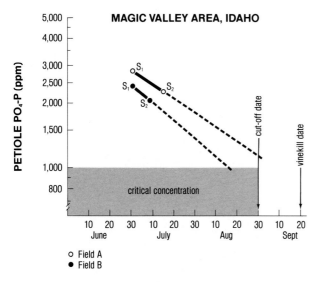

Figure 22. Two examples illustrating Idaho guidelines for estimating future petiole phosphate-phosphorus concentrations. Phosphate concentrations from two initial petiole samples (S_1 and S_2) are plotted on a semilogarithmic graph. The broken lines are extrapolated by drawing a straight line between S_1 and S_2 and predict a drop below the critical concentration (1,000 ppm) in field B before the cut-off date. Additional phosphorus fertilizer would be needed in field B, but not in field A.

- Avoid using chemical vine killers during cool, wet weather.
- Avoid uncovering or damaging tubers during mechanical operations.
- Make sure soil moisture is adequate (at least 50% of field capacity).

Start vinekill 2 or 3 weeks before harvest. Timing depends on cultivar, growing conditions, and time of year. Lush vine growth takes longer to kill; split applications of vine killing agents may be beneficial.

Harvest. Most potatoes are harvested by two-row harvesters directly into bulk trucks, which carry the tubers to a packinghouse, processor, or storage facility. Where soil is free of rocks and clods, windrowers are frequently used, allowing the harvest of four to six rows at a time.

Prevention of bruising is one of the most important considerations in a well-managed harvest operation. Blackspot and shatter bruise can seriously affect marketable yield if precautions are not taken to reduce them. Several factors are important in controlling bruising.

Soil Moisture. Proper soil moisture at harvest helps reduce bruising. The water content of tubers affects their sensitivity to bruising. If their water content is low, they are sensitive to blackspot bruise; if it is high they are sensitive to shatter bruise. The right combination of moisture and temperature helps to minimize both kinds of bruising. Soil moisture of 60% to 80% of field capacity is generally recommended.

Soil Temperature. The lower the soil temperature, the more susceptible tubers are to shatter bruising, especially when the temperature is below 50°F (10°C). Whenever possible, harvest when the soil temperature is above 50°F. Soils are warmest between 11 a.m. and 11 p.m. Harvest during these hours if soil temperature is likely to fall below 50°F at night.

Equipment Operation. Follow the recommendations for harvester operation in your area to reduce bruise injury. Use equipment that is in good repair, correctly adjusted, and run by an experienced operator familiar with local conditions. Important considerations are digger blade depth; reduction of "spill-out" losses at the digger blade; apron pitch, speed, and agitation; travel speed; drop heights; and controlling "undersweep." Operate the primary chain at 1.1 times the travel speed (1.5 feet per second for each 1 mile per hour of tractor speed). Operate the other harvester chains at 0.6 to 0.8 times the travel speed (1 to 1.2 feet per second for each 1 mile per hour). Keep chains filled so the tubers have little space to bounce around. Use padding on equipment wherever bruising might occur. When unloading at storage facilities, keep dropping heights to a minimum; use straw-filled bags on baffle boards; use mats underneath hoppers to cushion spilled tubers.

Tests are available to rapidly assess the degree of bruising. They can be used to determine whether you need to modify your equipment or handling procedures to reduce bruising.

Sprout Inhibitors. Apply sprout inhibitors to fresh market or processing potato tubers that are to be stored for more than 2 or 3 months. Low storage temperature cannot be used to prevent sprouting without undesirable accumulation of sugars. Foliar applications of sprout inhibitor are used on field-stored potatoes. Two types of sprout inhibitors are used: one, maleic hydrazide, is applied to potatoes while they are still actively growing; the other, chlorpropham (CIPC), is applied through the ventilation system in storage. Never apply these sprout inhibitors to seed potatoes.

If you are using maleic hydrazide, timing of application is critical. If applied too early, yields are reduced and injury may result; if applied too late it will not be translocated into the tubers, and sprouting will not be inhibited. Apply maleic hydrazide when all the tubers you intend to market are at least 1½ to 2 inches in diameter, when the plants are in the third (tertiary) bloom, or when lower leaves turn yellow. Do not apply the chemical to vines that are stressed by low water, frost, or disease. Some maleic hydrazide labels prohibit its use on irrigated potatoes in the western states.

Chlorpropham (CIPC) can be applied to stored tubers either as an aerosol spray or through the ventilation system. Apply CIPC after the curing period; use specially trained operators, and keep the storage closed for at least 48 hours after treatment. Avoid using CIPC in storage areas that are not equipped with forced air ventilation. Avoid handling tubers after treatment with chlorpropham; they do not heal (suberize) properly. Avoid storing seed tubers in a shed that has been treated with CIPC. If you must use such sheds for seed tubers, wait at least 6 months after CIPC treatment.

Storage. A large part of the crop in most growing areas is stored for fresh market or processing during the winter and spring. Russet Burbank is the main cultivar stored, and all seed tubers are stored. Designs can vary, but most storage facilities have controls for temperature, humidity, and ventilation. Ventilation is essential during storage. It removes field heat, excess moisture that may condense on colder tubers, and carbon dioxide and heat produced by respiration; at the same time it helps provide even temperature and humidity within the storage area and oxygen to support tuber respiration. Uniform airflow throughout the pile is important. For storage requirements in your area, consult local extension personnel or other experts.

The storage period consists of three phases: curing, holding, and warming.

Curing. During the first part of storage, hold tubers at a temperature of 50° to 55°F (10° to 13°C) with relative humidity above 95%. These conditions favor rapid suberization of any bruises or cuts incurred during harvest, and allow the skin of immature tubers to develop. Both of these processes increase the resistance of tubers to decay. Use a curing temperature of 45°F (7.3°C) for tubers that will be processed into french fries or dehydrated products, or sold for fresh market. Use temperatures of about 50°F (10°C) for chipping cultivars to keep sugar levels down. Hold tubers under curing conditions for a minimum of 2 weeks. When there is increased risk of decay, as with tubers injured by frost or tubers exposed to late blight or excessively wet conditions during harvest, store affected lots separately. Cool and dry them as quickly as possible with high flows of non-humidified air.

Holding. For most of the storage time, hold tubers at the lowest temperature possible without affecting market quality. The following holding temperatures are recommended:

- chipping, 50° to 55°F (10° to 13°C);
- french fries, 45° to 50°F (7.3° to 10°C);
- fresh market, 40° to 45°F (4.4° to 7.3°C);
- seed, 35° to 40°F (1.7° to 4.4°C).

Maintain high humidity to keep tubers from drying and to avoid development of pressure bruise.

Warming. If holding temperatures were lower, letting tubers warm up to 50°F (10°C) before removing them from storage will reduce bruising. Allow the heat of respiration to warm tubers. Do not use warm air; condensation may occur on cold tubers, creating conditions that favor decay. Be sure to maintain humidity to keep tuber water content up. Tubers with lower water content are more susceptible to blackspot bruising. If excessive sugars have accumulated in tubers to be used for processing, warming above 50°F for 3 weeks may reduce their sugar to acceptable levels. Before cutting seed tubers, warm them at least 10 days at 50° to 55°F (10° to 13°C) to increase their wound-healing ability.

Pesticides

Anyone working with pesticides should know the identity of the chemicals being used; their potential for injury to humans, livestock, crop plants, or the environment, necessary safety precautions, and what to do in case of an emergency. Know the target pest for each application and which nontarget species, especially natural enemies and pollinators, may be affected.

Properly used, pesticides can provide economical protection from pests that otherwise would cause significant losses. In many situations, they are the only feasible means of control. Careless or excessive use of pesticides, however, can result in poor control, crop damage, higher expenses, and hazards to health and the environment. In an IPM program, pesticides are used only when field monitoring indicates they are needed to prevent losses.

In choosing a pesticide, consider not only its effect on the target pest, but also the effects it may have on other pests, natural enemies, and crop plants. Effects may vary according to formulation, rate, and application method. Consult your local extension agent, farm advisor, or other experts for current information on new materials and methods before making a choice.

Make sure each pesticide treatment is suited to the appropriate stage of the target pest and is timed to coincide properly with crop growth, weather conditions, and cultural operations. If you can get the control you need with fewer treatments, you will reduce costs and potential hazards. Also, the chance of such side effects as pest resurgence, secondary pest outbreaks, and pesticide resistance will be lessened.

Always READ THE LABEL carefully before using any pesticide. Follow directions and observe suggested safety precautions. Be aware of changes in labels of the pesticides you use.

Pesticide Resistance. Some insect and mite pests of potatoes have developed resistance to certain pesticides; they are able to survive applications that once killed most individuals of the same species (Table 11). At first, it may

be possible to control resistant pests by increasing rates, but higher rates and more frequent exposures increase the chance that resistant individuals will increase as a proportion of the total population (Figure 23). When that happens, the pesticide involved will no longer be economically useful. You may then switch to a new material. The search for new pesticides is complicated by the fact that species resistant to one material often develop *cross-resistance* to others, even to materials not in the same chemical class. Resistance to new pesticides may then develop much more quickly than it did with the original one. For example, insect pests that have developed resistance to chlorinated hydrocarbons such as DDT are more likely to become resistant to pyrethroids. Also, pesticide resistance can develop in a pest not intended as the target of applications.

Pest Resurgence and Secondary Outbreaks. Pesticides that kill or disturb natural enemies may cause pest resurgence or outbreaks of secondary pests. These problems are most common with the use of insecticides.

Pest resurgence occurs when a pesticide destroys both the target pest and its natural enemies. Because the natural enemies depend on the pest for food, they take longer to build up to their former numbers. On the other hand, pests that survive treatment or that move into the field later can breed without the restraint of natural enemies, sometimes increasing to greater numbers than existed before treatment.

Pesticide applications may also cause damaging increases in pests that were not targets. Such increases are called secondary outbreaks, and usually result from destruction of natural enemies that controlled the secondary pest before the application (Figure 24). For instance, secondary outbreaks of spider mites may occur in potatoes when natural enemies are destroyed by insecticides applied for aphids or other insect pests. Likewise, application of carbaryl to control chewing insects often results in increased aphid populations.

Accelerated Breakdown. Breakdown by certain soil microbes is the primary means by which many pesticides, including soil-applied herbicides, are removed from the environment. Repeated applications of certain herbicides or insecticides may encourage populations of these microbes to increase. When this happens, the pesticides are broken down so fast that they become either less effective or useless. In some fields in the Northwest, S-ethyl dipropylthiocarbamate (EPTC) has lost its effectiveness due to the buildup of microbes that degrade it. Avoid repeated applications of EPTC, and use the lowest rates necessary for weed control. Problems with accelerated breakdown have also been reported for carbofuran.

Crop Injury. Crop injury (phytotoxicity) from pesticides can result from an improper application method, poor

Table 11. Pesticides for which Resistance Has Been Reported in Insects and Mites of Potato in the United States. Resistance in a particular species does not necessarily occur uniformly in all areas.

Pest Species	Pesticide
Colorado potato beetle[a]	carbamates, chlorinated hydrocarbons, organophosphates, and fenvalerate.
Green peach aphid	Many pesticides, including carbamates, chlorinated hydrocarbons, organophosphates, and pyrethroids.
Cabbage looper	chlorinated hydrocarbons and organophosphates.
Twospotted spider mite	chlorinated hydrocarbons and organophosphates.

a. Significant insecticide resistance is not yet known to occur in the western states.

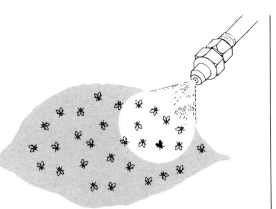

Some individuals in a pest population have
genetic traits that allow them to survive a
pesticide application.

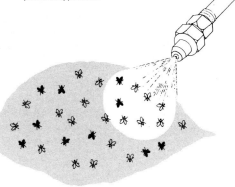

A proportion of the survivors' offspring
inherit the resistance traits. At the next
spraying these resistant individuals
will survive.

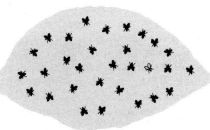

If pesticides are applied frequently, the
pest population will soon consist mostly
of resistant individuals.

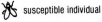 susceptible individual

resistant individual

Figure 23. Pest populations develop resistance to pesticides
through genetic selection.

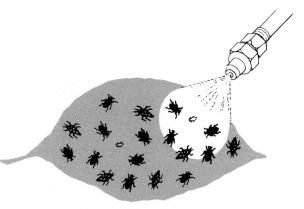

A pesticide applied to control pest A also
kills natural enemies that are controlling
pest B.

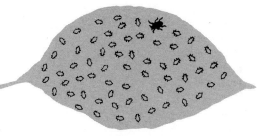

Released from the control exerted by
natural enemies, pest B builds up to economically damaging levels

 pest A

pest B

 natural enemy

Figure 24. Secondary outbreaks of insects and mites are often
caused by destruction of natural enemies.

timing of application, excessive rates, drift, and residues in soil or water. Most common problems are due to herbicides. Some potato herbicides may injure certain cultivars; be sure to follow label restrictions regarding cultivars. Metribuzin may injure potato plants if applied within 3 days following cloudy, cool weather. Maleic hydrazide may cause injury if improperly applied.

If you plan to follow potatoes with a rotation crop, make sure the herbicides you use will not leave residues potentially harmful to that crop. Before planting potatoes, be sure no harmful residues are left over from the rotation crop.

Hazards to Human Health. Some pesticides used in potatoes are hazardous to humans. The applicator is most at risk; field personnel, irrigators, and others who enter the field may also be exposed. Pesticides that drift into roadways, yards, or other areas outside the field may also cause health problems. Some pesticides may contaminate groundwater supplies. Aldicarb use is prohibited in some areas of the East and Midwest because of groundwater contamination. It is not recommended in areas of the West that have shallow water tables and sandy, acidic soils, because the potential for groundwater contamination is too great. Read and follow the safety precautions on pesticide labels and consult the your extension agent, farm advisor, or other experts for guidelines on how to use pesticides safely. Several safety publications are listed in the References.

Before applying a pesticide, learn what emergency measures can be used safely by field personnel. Make sure application equipment is in good condition and has the appropriate safety features for the material you are using. Always wear appropriate protective clothing when handling or applying pesticides. Dispose of containers according to label directions.

Hazards to Wildlife. Pesticides applied to potatoes can contaminate water and kill fish, birds, or other animals if contaminated water drains into streams or other bodies of water. Do not apply pesticides if irrigation water may run off into streams, and do not let pesticides drift onto water. Keep abreast of regulations designed to protect wildlife from pesticide contamination.

Insects and Related Pests

Insects injure potatoes by transmitting pathogens and by feeding on tubers, seed pieces, leaves, and stems. The most damaging insect pest in the western United States is the green peach aphid, which transmits the virus causing leafroll and tuber net necrosis as well as other potato viruses. In commercial potatoes, aphid monitoring and control are intended to keep net necrosis within acceptable levels in harvested tubers. Management of aphids is especially important in Russet Burbank potatoes, which are highly susceptible to net necrosis. Aphid control is critical in seed potatoes, regardless of variety, because of the low tolerance for leafroll and other viral diseases. Although insecticides are needed in many cases, aphid control is only one aspect of virus management; the use of certified seed, the elimination of culls and volunteer plants, and other cultural practices are also essential.

Pest species that attack tubers or seed pieces occur throughout the region, but their importance varies from one area to another. Wireworms are especially damaging in Washington, Oregon, Utah, and Idaho. Wireworm monitoring requires keeping records of damage to potatoes and other crops; soil-applied insecticides are needed occasionally for control. In central and southern California, the potato tuberworm is the most important insect pest; a monitoring program based on pheromone traps is available for timing sprays. Other tuber pests, including flea beetles, symphylans, cutworms, and seedcorn maggot, occur either sporadically or only in limited areas. Pests that destroy foliage are most damaging during the midseason period when tubers are developing. The Colorado potato beetle, which feeds on foliage in both the larval and adult stages, is a major pest in Idaho, Montana, Oregon, Washington, Wyoming, and portions of Colorado. Certain systemic insecticides applied against aphids partially control the beetle, although foliar insecticide treatments may also be needed, depending on the length of the season and the number of pest generations.

Other foliage pests, such as flea beetles, loopers, cutworms, and spider mites, are widespread but are rarely damaging enough to warrant special control measures. Insecticides applied for green peach aphid or Colorado potato beetle often control these minor pests as well.

The potato psyllid, which damages plants by injecting a toxin during feeding, is an important but sporadic pest of potatoes in Arizona, New Mexico, Utah, Colorado, and Wyoming. Monitoring is the same as for green peach aphid; foliar sprays are needed for control in some seasons.

Some insects found in potato fields are beneficial rather than harmful. Parasitic wasps such as *Hyposoter exiguae* attack loopers, armyworms, and other caterpillars. Parasitic flies known as tachinid flies destroy Colorado potato beetles, and other tachinid fly species attack caterpillars. Predators such as the twospotted stink bug, ladybird beetles, and damsel bugs are also common in potato fields. Although natural enemies seldom keep major pest insects from reaching damaging levels in potato fields, they can be important in limiting pest populations on other hosts, thus reducing the numbers that invade potato fields. Natural enemies also help to keep minor pests such as loopers from becoming more damaging.

Monitoring

Visit each field at least once a week throughout the season to watch for pest activity and to make sure the crop is developing normally. As soon as shoots start emerging, begin walking the field regularly to look for blank spots, weak shoots, and infestations on young plants. If you find weak areas or damaged shoots, check seed pieces and look in the soil for soil-dwelling pests such as wireworms, seedcorn maggot, and symphylans. The most important monitoring activity, especially where Russet Burbanks are grown, is sampling for aphids; the procedure is described in the green peach aphid section of this chapter. Keep alert for outbreaks of psyllids, spider mites, Colorado potato beetles, and other foliage pests.

Other monitoring activities are needed only in certain situations. In newly cultivated land, in fields planted to potatoes following weedy alfalfa or pasture, and in places where wireworm injury was noted in previous crops, you may need to collect soil samples before planting and sift them to look for wireworms. In those parts of California where the potato tuberworm occurs, set out pheromone traps and check them twice a week to monitor activity of the tuberworm adults.

SUCKING INSECTS THAT DAMAGE POTATOES

Insects such as aphids and psyllids use their specialized mouthparts to puncture the host plant and withdraw sap from the phloem or food-conducting tissue. Aphids injure potatoes primarily by spreading viruses such as leafroll. Occasionally the number of aphids may be high enough that their feeding injures potato plants directly.

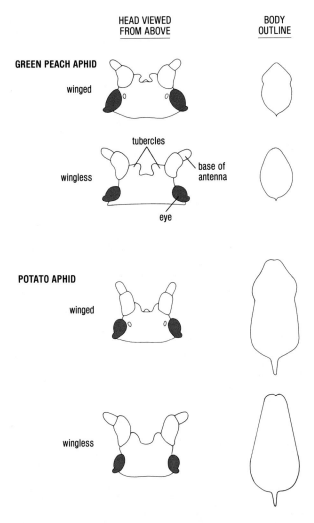

Figure 25. The tubercles between the bases of green peach aphid antennae converge; the body is short and oval in outline. The tubercles of the potato aphid slope outward toward the antennae, and the body is elongate.

Psyllid nymphs inject a toxin while feeding that causes psyllid yellows.

Green Peach Aphid
Myzus persicae

The green peach aphid (GPA) is a major pest everywhere potatoes are grown. It is the vector of potato leafroll virus PLRV, the pathogen causing leafroll and tuber net necrosis. Green peach aphid also transmits other viruses. Control is based on the use of systemic, soil-applied insecticides early in the season or foliar sprays later. Monitoring methods are available for scheduling foliar applications in commercial potatoes in some areas. Practices that reduce overwintering populations have reduced some infestations in potato. Seed potatoes have an especially low tolerance for green peach aphid because the incidence of potato leafroll virus must be very low for seed to be certified.

Description

Like most aphids, green peach aphid has both winged and wingless forms. Wingless forms are teardrop shaped and usually light green, although some populations, especially in late summer, include many individuals that are pink. In winged forms, the head and thorax are brown or black, and the abdomen is green with a darker patch on the back. Immature aphids that are developing into winged forms have wing pads.

Green peach aphid is by far the most common aphid on potatoes in the western United States—more than 90% of the aphids found on potatoes in the region are green peach aphid. The next most common species is the potato aphid; others are found only occasionally. You can distinguish the green peach aphid from potato and other aphids by size, body shape, and the shape of the antennal tubercles (Figure 25).

Seasonal Development

The green peach aphid has a complex life cycle with 10 to 25 generations a year (Figure 26). Most generations consist only of females that produce live nymphs without mating. Sexual reproduction is absent altogether in areas with mild winters (Figure 27). The population spreads from one host to another all year as new hosts become available and others dry out or are destroyed by frost. The green peach aphid feeds on several hundred plant species, including many crops, weeds, and ornamentals.

In areas with cold winters, a generation of sexual forms appears in the fall and eggs are laid on a winter host, usually a peach tree. Overwintering eggs hatch in the spring, producing a generation of wingless females called stem mothers. The stem mothers feed on the buds and young leaves of the winter host; each one produces 100 to 200 wingless females, beginning a series of generations on the winter host.

Starting in the third spring generation, some nymphs develop into winged migrants and fly to other hosts to begin a series of summer generations. When a migrant reaches a new host plant, it usually remains only long enough to produce a few nymphs, then moves to yet another plant. Most flights by migrants are short, so infestations usually spread gradually from the winter host. Migrants can, however, be carried long distances by wind, so even plants far downwind may be infested. Early spring hosts include mustards and related weeds in and around orchards.

Each nymph deposited by a spring migrant has the potential to start, by asexual reproduction, a new colony of nonmigrant, nonwinged aphids. After one or more generations, the colony begins to produce a proportion of winged individuals (summer migrants) in each generation. The proportion of migrants to nonwinged adults increases in each generation as colonies become crowded and as host plants dry out or otherwise become less suitable for the aphids. A nymph on a summer host develops to maturity and begins producing a new generation of nymphs in as few as 6 days. In cooler weather, development may take 2 weeks or more.

Winged aphids that colonize potato may come directly from peach trees or other winter hosts, but more commonly they reach the field from intervening hosts such as mustards, nightshades, groundcherries, or ornamentals. Infestations usually advance through the field in a downwind direction. Plants at the edge of the field are infested first and generally remain more heavily infested throughout the season than plants in the center of the field. The aphid population is usually concentrated on the lower third of the plants. Upper leaves may be infested during long periods of cool, cloudy weather or when aphids on lower leaves become crowded.

In most potato growing areas, the green peach aphid population reaches a peak in the latter half of the season, then declines due to the senescence of host plants, unfavorable temperatures, activity of natural enemies, and emigration of winged forms. Where the season is too short for the population to reach the decline phase, it simply continues increasing through the season.

In late summer or fall, a generation of winged migrants appears that includes both males and females. The females leave the summer host where they developed and fly out at random in search of a winter host. Upon reaching a winter host, they deposit nymphs that become wingless females. Male migrants that reach the winter host mate with these wingless females, which then lay about 5 to 15 overwintering eggs on buds that will open in the spring. Eggs can survive temperatures as low as −50°F (−46°C).

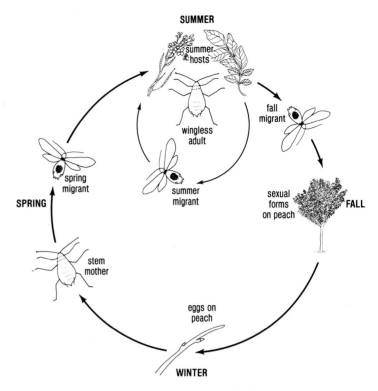

Figure 26. Seasonal population cycle of green peach aphid in areas with cold winters.

Figure 27. In the areas of California and Arizona below the dotted line, the green peach aphid develops all year in a continuous series of generations with no egg stage. In the rest of the western United States, sexual forms appear in the fall and lay eggs on a winter host.

The predominant winter host in the western states is peach, *Prunus persica*, although green peach aphid overwinters on other *Prunus* species in some cases. In Idaho, green peach aphid occasionally overwinters on apricot. In Washington, green peach aphid lays overwintering eggs on various *Prunus* species, but winter survival has been demonstrated only on peach and nectarine. Winter survival on hosts other than peach, including wild *Prunus* species, has been suggested as a possible source of infestations in areas such as the Tule Lake Basin of California and the San Luis Valley of Colorado, where green peach aphid is common even though peach trees are rare. Green peach aphid may survive on American plum, *Prunus americana*, in the San Luis Valley, but not on choke cherry, *Prunus virginiana*. However, choke cherry is a common winter host for green peach aphid in the eastern United States.

The eggs require a certain period of chilling and exposure to water to develop. After the chilling requirement is reached, eggs hatch in response to warm weather. At low elevations, eggs may hatch as early as January or February; at higher elevations or further north, they may not hatch until late April.

In areas with mild winters (Figure 27), green peach aphid has no sexual phase; there are no males and no eggs are produced. The population continues moving from one host to another during winter in a series of asexual generations. Even in colder regions, green peach aphid may continue development all winter in greenhouses and other situations where host plants are available and where temperatures remain favorable. Bedding plants, such as vegetables and cut flowers, that are produced in greenhouses and set out in the spring are often infested and can be an important source of green peach aphids that later move to potato. Populations may also overwinter in protected places around heated buildings, along canals, and around springs.

Damage and Relation to Viral Diseases

Injury to potato by green peach aphid is due mostly to transmission of potato leafroll virus and other potato viruses. Where a reservoir of the virus exists, a relatively small green peach aphid population can be damaging, especially to seed potatoes. Severe infestations can reduce growth or cause other symptoms by draining plant fluids, but aphids are rarely numerous enough for such injury to be significant in potato production.

Transmission of Potato Leafroll Virus. To transmit potato leafroll virus to a healthy plant, an aphid must first acquire the virus by feeding on a potato plant that is already infected. A period of time called the *latent period* is required before the aphid can transmit potato leafroll virus to a new host plant. The latent period is at least a day, frequently longer. During the latent period, the virus passes through the aphid's gut, into its blood, and finally into its salivary glands. When the aphid moves to a new host plant and begins feeding, virus particles move with the salivary fluid into the plant. About 30 minutes of feeding is required to produce an infection. Once an aphid acquires the virus, it usually remains infectious for life. The virus does not pass from adult aphids to their nymphs or eggs.

The spread of potato leafroll virus to a potato field or from one field to another is due to the movement of winged migrants. Migrants typically live for up to 3 weeks, although most flight activity occurs in the first few days. Wingless aphids can also carry the virus, either by passing directly from leaves of an infected plant to those of adjacent plants, or by crawling over the ground to nearby plants. Both winged and wingless aphids, therefore, can spread potato leafroll virus within fields; adjacent plants in the same row as an infected plant are the most vulnerable to virus transmission.

Transmission of Other Viruses. The green peach aphid is one of several aphids that act as vectors of alfalfa mosaic virus (calico) and potato viruses A and Y. These viruses are called *styletborne* or *nonpersistent*, and do not circulate within the aphid's body as does potato leafroll virus; they are carried on the mouthparts and transmission occurs rapidly when contaminated aphids probe host plants. Because the virus particles are quickly lost from the mouthparts, aphids can transmit them only once or a few times, rather than repeatedly over a long period as in the case of potato leafroll virus.

Alfalfa mosaic virus is transmitted mostly by aphids carried by wind from alfalfa fields. Although green peach aphid can carry the virus, aphid species that are more common on alfalfa are the major vectors; these include the pea aphid, *Acyrthosiphon pisum*, and the blue alfalfa aphid, *A. kondoi*. Viruses Y and A are generally carried by aphids from infected potato plants, although these viruses can also infect nightshades and other weeds. In commercial potatoes, yield loss caused by these viruses is generally not great enough to justify aphid control measures in addition to those required for preventing leafroll. Transmission of nonleafroll viruses is damaging, however, in seed potato fields; special control measures may be needed in some cases.

Management in Commercial Potatoes

The impact of a green peach aphid population in commercial potatoes depends on the number of infected potato plants that serve as reservoirs of potato leafroll virus. If the number of reservoir plants is high, even a few aphids can be damaging. If infected plants are absent, aphids cause little damage unless heavy infestations stress the plants. Since it is seldom possible to be certain that reservoir plants are not present, it is always best to prevent large aphid populations from building up, especially in Russet Burbanks and other cultivars susceptible to tuber net necrosis. Russet Burbank tubers grown in the San Luis Valley of Colorado do not develop net necrosis.

Transmission of potato leafroll virus is most likely to result in net necrosis of tubers if it occurs while tubers are expanding, so this is the stage of growth at which aphid control is most important. However, young plants are highly susceptible to potato leafroll virus infection, so a few virus-carrying aphids during early growth can significantly increase the incidence of potato leafroll virus in the field. Secondary spread by aphid populations that develop later in the season can then result in high amounts of net necrosis in Russet Burbank tubers. Although insecticides are often needed, they are effective only in combination with other practices, including the use of certified seed, the elimination of culls and volunteer plants, roguing of infected plants, weed control, and removal or treatment of aphid overwintering hosts.

Sanitation to Reduce Virus Reservoir. Field sanitation and the use of certified seed make up the first line of defense against leafroll. In the absence of these measures, insecticides probably will not prevent high aphid populations from transmitting damaging levels of potato leafroll virus. The use of infected seed pieces increases the likelihood of leafroll not only in the current year's crop, but also in the next season's, because volunteers may sprout from tubers produced by infected plants and left in the field after harvest.

A suitable sprout inhibitor applied to the vines before harvest can reduce emergence of infected volunteers that could serve as a reservoir of virus for the following year's crop. Although it may be more expensive than postharvest treatment, an in-field treatment with inhibitor may be worthwhile in cases where the chance of leafroll infection is high due to a large green peach aphid population or the presence of a large potato leafroll virus reservoir. Follow label directions carefully in timing applications of sprout inhibitor; applying inhibitor too early reduces yield, while applying it too late reduces effectiveness.

Weed control is also an important part of sanitation for reducing green peach aphid populations and potato leafroll virus transmission. Follow recommended weed control procedures to keep fields as free as possible of green peach aphid hosts such as mustards, nightshades, and groundcherries. If possible, eliminate stands of these weeds in fencerows, ditches, and other uncultivated areas near the field.

Treatment of Winter Hosts. Where green peach aphid overwinters in the egg stage, this is the most vulnerable

part of its life cycle. Peach orchards are usually treated in the spring for peach twig borer or other pests; these treatments are also effective for green peach aphid if they are applied after eggs hatch but before winged migrants develop. Also, normal pruning in orchards destroys many green peach aphid eggs. When orchards are abandoned or damaged by frost, however, they may not be pruned or treated for peach pests, and a special treatment for green peach aphid may be worthwhile. Use a spray of dormant oil plus a suitable insecticide, or apply an insecticide as a delayed dormant spray. The same treatments are recommended for backyard peach trees.

Orchards where the cover crop is a green peach aphid host produce a much larger aphid population than those in which weeds are controlled or the cover crop is a grass. If possible, try to keep peach orchards near potato fields free of weed hosts.

Because untended peach trees around homes and volunteer trees along roadsides are often unpruned and surrounded by weeds, these trees may produce exceptionally large numbers of aphids. Removal or treatment of such trees growing near potato fields is highly desirable. Bedding plants can be sources of green peach aphid. Petunias, green peppers, eggplants, cole crops, and occasionally tomatoes are green peach aphid hosts. Inspect bedding plants for aphids before moving them outdoors; treat them with a suitable insecticide if they are found to be infested.

Community Aphid Control Programs. Because green peach aphid has a large number of hosts and completes much of its life cycle outside of potato fields, some practices are effective only on a communitywide basis. In some areas this is the only way to prevent virus-carrying aphid populations from moving into seed fields. Model programs established in southeastern Idaho and the San Luis Valley in Colorado have demonstrated that areawide efforts can reduce the incidence of leafroll significantly. The most important practices that require community coordination are the removal or treatment of peach and apricot trees and treatment of greenhouse stock to eliminate green peach aphid from bedding plants such as petunias, green peppers, and eggplants set out in the spring. More background on the Idaho program is available in the publication *Management of Potato Insects in the Western States*, listed in the References.

Monitoring. Two recommended methods for monitoring green peach aphid are leaf sampling and pan trapping. Leaf sampling is the best method for timing foliar treatments in specific fields. Pan traps are useful for monitoring aphid activity on an areawide basis, but they do not provide a reliable basis for scheduling insecticide applications. By the time winged green peach aphids appear in pan traps, a low level of infestation may already be present on potato plants. There is no specific method for deciding whether a systemic insecticide is needed early in the season; make your decisions by evaluating the virus reservoir, the threat of net necrosis, and the potential early season aphid population based on information from previous seasons. Methods recommended for leaf sampling vary among the different areas in the western United States, but the following directions are applicable in most cases.

1. Where an orchard, community, or other potential aphid source is nearby, include the side of the field closest to this source in the sample. Start in one corner of the field and move toward the center as you pick sample leaves (Figure 28). In large fields, move at least 300 to 400 feet (100 to 130 m) into the field before turning back toward the edge. In smaller fields, move to the center before turning back. If potential aphid sources are not nearby, sample all four quadrants of each field (Figure 29).
2. Pick one leaf from each plant sampled. Remember that potato leaves are compound, each with several leaflets; count aphids on all leaflets of each sampled leaf.
3. Pick sample leaves from the lower third of the plants; this is where green peach aphid is concentrated.
4. Step at least two paces between samples.

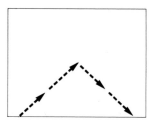

Figure 28. If potential aphid sources are nearby, sample the side of the field closest to the source (e.g., an orchard or a community). Start near one corner of the field and move toward the center as you pick sample leaves.

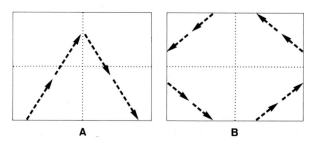

Figure 29. For a thorough aphid sampling, include all four quadrants of each field by following a pattern such as in A or B. Check field edges in a separate set of samples.

5. Depending on the treatment threshold for your area, either count all wingless aphids on the underside of the sample leaf or simply record whether the leaf is infested. If you are counting aphids, you can collect several leaves before stopping to count, because aphids do not readily move off the leaves.

The following treatment thresholds are suggested for various parts of the western United States:

Idaho. In southwestern Idaho, treat when there are more than 40 wingless green peach aphids per random sample of 50 leaves in 2 consecutive weeks before August 1. In central and eastern Idaho, treat if you find more than 10 wingless green peach aphids per random sample of 50 leaves in 2 consecutive weeks before August 1. Treatment thresholds for Idaho apply only to fields where certified seed is used and only to fields that are not near peach orchards or near unsprayed garden peach trees and are not adjacent to potato fields with excess leafroll. They do not apply to fields where significant numbers of volunteer potato plants are present. Treatments are not needed if aphids appear after August 1.

California. In the Tule Lake Basin of California, treat when 5% of the sample leaves are infested with green peach aphid. The minimum sample is 100 leaves; pick only the lower-most leaves on each sample plant. There is no established threshold in other parts of California, because aphids rarely cause economic damage to cultivars grown for commercial production in these areas.

Utah. In Utah, follow guidelines for southwestern Idaho.

Other States. There are no established treatment thresholds for green peach aphid on potatoes in other western states.

General Guidelines. Check potato plants at the edges of the field separately from regular leaf sampling; if an infestation builds up at the edges, it may be worthwhile to apply a strip treatment of foliar insecticide there. In fields where systemic insecticides are applied at planting, there is often an untreated section at the end of the rows due to the space required for the planter to turn around; this section may be infested earlier than the rest of the field and require special attention in monitoring. Also, watch for green peach aphid on mustards, nightshades, ground-cherries, and other weed hosts in and around the field.

Pan traps, though not suitable for scheduling treatments, help to identify periods when aphids are abundant and when field monitoring is most needed. A pan trap is a yellow container of water, such as a plastic dishpan, filled to within 3 inches of the top; aphids attracted by the color are trapped when they land in the water. A mold inhibitor added to the water helps to keep

Most green peach aphids are light green; some may be pink. The shape of the tubercles at the base of the antennae distinguish them from other aphids found on potatoes (see Figure 25).

Green peach aphid eggs are laid on or near buds of peach trees or other overwintering hosts. The eggs are dark green at first, then turn shiny black. One black egg can be seen here between the twig and axillary bud.

The winged adult green peach aphid has a dark brown or black head and thorax, and a green abdomen with dark patches. Adults may also be wingless.

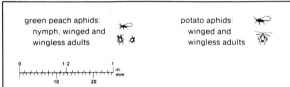

APHIDS

green peach aphids:
nymph, winged and
wingless adults

potato aphids:
winged and
wingless adults

0   1/2   1 in.
                  mm
   10   20

Pan traps are used to determine when aphid populations are building up and when field monitoring is necessary. The yellow color attracts aphids and other insects, which are trapped in the water.

aphid specimens in good condition. Each trap is set on a background of contrasting darker material on the ground, near the crop but not surrounded by vegetation. Pan traps must be checked at least once a week—often enough so the water does not evaporate between checks.

Several aphid species other than green peach aphid, including potato aphid and some that do not feed on potato, may appear in pan traps. Aphids must be removed from traps with a forceps or strainer, placed in alcohol, and taken to a lab where they can be examined under a low-power microscope for proper identification. Because of the time required, pan traps are usually operated by local extension agents, farm advisors, growers' cooperatives or other agencies which report the results to growers. Pan trapping is generally not worthwhile for the individual grower.

Chemical Control. Insecticides for aphid control in potato include soil-applied systemic materials as well as foliar sprays. Systemics, applied at planting or as side-dressing, control aphids for about 6 to 8 weeks or longer after application. Systemic insecticides also control other pests, including Colorado potato beetle and some nematodes. However, soil-applied systemics are not recommended in certain areas such as California and the San Luis Valley of Colorado, where net necrosis is not a threat or aphids become active after systemics have worn off. Foliar sprays must be used if control is needed after a soil application has ceased to provide adequate control, and in areas where soil-applied systemics are not recommended.

Systemic insecticides prevent aphid populations from building up, but they do not kill aphids quickly enough to prevent winged migrants from transmitting viruses. If large numbers of migrants are flying in from virus reservoirs, the incidence of leafroll or other viruses may be high despite treatment with systemics. If there are infected plants within the field, winged aphids from outside may live long enough to acquire virus from an infected plant and transmit it to a healthy one, even though both plants contain a systemic insecticide.

In areas where the virus reservoir is reduced through sanitation and use of certified seed, low aphid populations during the first few weeks—the period when systemics are effective—are less likely to be damaging. However, because young plants are highly susceptible to virus infection, presence of systemic insecticide during early development is considered important for avoiding unacceptable levels of net necrosis in most Russet Burbank growing areas. Although systemics are usually applied at planting, postemergence side-dressing may be recommended instead if slow emergence is expected (for example, when potatoes are planted early in the spring). Otherwise, much of the systemic material may be broken down before plants emerge. If a side-dressing is needed, apply it after 75% of the plants have emerged but before plants reach 6 to 8 inches (15 to 20 cm) in height; later application may damage stolons. Follow the application with an irrigation.

Where special risk factors are not present and where aphid populations do not reach damaging levels until the latter half of the season, you can rely on foliar sprays for aphid control. In these situations, one or two properly timed applications of a suitable insecticide usually provide adequate control. Take into account the results of local pan-trapping programs and any information on aphid numbers in other crops to help decide when the level of aphid activity warrants treating potatoes. To schedule foliar treatments, follow the directions for leaf sampling earlier in this chapter and apply insecticide according to current local recommendations. In some cases, it may be necessary to treat a potato crop to protect other potato fields nearby. For example, a properly timed spray can prevent the movement of winged aphids from varieties such as Norgold or Kennebec, which are less susceptible to net necrosis and can tolerate higher green peach aphid populations, to a field of highly susceptible Russet Burbanks. Also, it may be necessary to treat commercial potatoes to prevent aphids from moving to nearby seed potato fields.

Management in Seed Potatoes

Because of the low tolerance for viruses in seed potatoes, more thorough aphid control and extra sanitation practices are needed in seed fields. Essential practices for reducing virus infections include roguing infected plants along with surrounding plants (Figure 30), harvesting before large aphid populations build up, elimination of volunteer potato plants, selection of seed for increase from the middle of the field, and control of aphids in nearby commercial fields. A single flight of aphids from a field with high incidence of virus can cause infections in a seed field that will exceed certification tolerances, even if systemic insecticides have been applied. It is easier to keep virus incidence below the tolerance levels required for seed certification in short-season areas where green peach aphid population development is delayed, host trees are scarce, and seed fields are isolated from commercial potatoes.

The same control methods are used for green peach aphid in seed as in commercial potatoes, but most areas have no specific treatment thresholds because the tolerance for aphids is so low. The treatment threshold used in Montana is given below. Preplant application of systemic insecticides is recommended in the seed-producing

Figure 30. If you find a plant with leafroll symptoms in a seed potato field, remove and destroy it and adjacent plants in the pattern shown here.

areas of Idaho, Oregon, Washington, and Montana. In some areas, such as the San Luis Valley of Colorado, it may be possible to wait until aphids appear before applying an aerial spray. Soil-applied systemic insecticides are not recommended in California or the San Luis Valley.

In Montana, if one aphid is found in a field, wait one week and sample again. Treat if aphids are still observed. Take a minimum of five samples for 40 acres, and one additional sample for each additional 10 acres. For each sample take 10 leaves from the lower third of the canopy and 10 leaves from the upper third.

As soon as potato leaves are present, start checking them for green peach aphid. As plants grow, keep checking the lower leaves, where green peach aphids are usually concentrated; aphids are often present even on older, yellowed leaves that are touching the ground. Be sure to check mustards, nightshades, and groundcherries, too; green peach aphid often appears on these weeds before moving to potato. Concentrate monitoring along the edges of the field, especially on the upwind side and in sections closest to peach trees, other green peach aphid hosts, or nearby communities. In some cases, it may be worthwhile to apply a spray only to the edge of the field to destroy an aphid population that might later move into the field.

In fields that have been treated with a systemic insecticide, watch for the appearance of nymphs as a signal that the insecticide is no longer controlling aphids. If you find only winged aphids and a few small, isolated nymphs, the treatment probably is still effective; winged migrants often live long enough in treated fields to deposit a few nymphs. However, if you find wingless aphids surrounded by a group of several nymphs, the aphids must have been feeding for several days and the insecticide is no longer effective.

Results of pan trapping can be useful in scheduling harvest of seed potatoes; if green peach aphid activity is increasing steadily, you may need to harvest or kill the vines before the number of migrants reaches such a level that the population in potato can no longer be suppressed adequately with sprays.

Potato Aphid
Macrosiphum euphorbiae

The potato aphid is the only aphid other than green peach aphid commonly found on potato in the western United States. It is generally less damaging because it transmits potato leafroll virus in a nonpersistent fashion, usually just the first time it feeds after acquiring the virus. It also transmits potato viruses Y and A in the same way. The potato aphid occurs throughout the region.

The potato aphid differs morphologically from green peach aphid in the characteristics shown in Figure 25.

Potato aphids are larger and more elongated than green peach aphids. The tubercles at the base of the antennae do not converge as they do in green peach aphids.

There are winged as well as wingless forms of potato aphid adults. They may be green or pink.

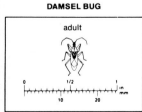

DAMSEL BUG

adult

Damsel bugs feed on the different stages of a variety of insects, including aphids, leafhoppers, and psyllids. Both the nymph (*right*) and adult (*left*) are predators.

The potato aphid is larger than the green peach aphid, the body is more elongate, and potato aphids do not have the distinctive converging antennal tubercles found on green peach aphids. Potato aphids may be green or pink. They feed mostly on upper leaves rather than on lower leaves where green peach aphids concentrate.

The life cycle of the potato aphid is similar to that of the green peach aphid. The most common winter host is rose. Aside from potato, other summer hosts include tomato, groundcherry, and nightshades. Nightshade is a preferred host.

In commercial potatoes, control of potato aphid is needed only in rare cases when extremely large populations stress plants by draining sap. Control may be needed in seed potato fields or in commercial fields in seed-growing areas to limit transmission of viruses. Insecticides for green peach aphid also control potato aphid, so special applications for potato aphid are rarely necessary.

Potato Psyllid
Paratrioza cockerelli

Psyllids are small, cicadalike insects about 1/12 inch (2 mm) long. They are related to aphids and leafhoppers and feed in the same way, sucking plant sap through needlelike mouthparts. The potato psyllid occurs throughout the western United States. Damaging infestations are most common in Arizona, New Mexico, Colorado, Wyoming, and southern Utah. Natural enemies of potato psyllid include minute pirate bugs, damsel bugs, and a *Tetrastichus* parasitic wasp. Outbreaks of psyllids are sometimes associated with the use of systemic carbamate insecticides, which are ineffective against

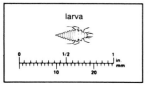

LACEWING

larva

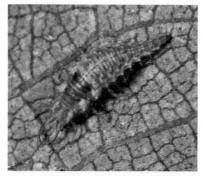

Green lacewing larvae eat large quantities of mites, aphids and other small insects, and insect eggs.

Potato psyllid eggs are borne on stalks at the margins of leaflets.

BILL CALLISON

Psyllid toxin stunts the growth of potato leaves and causes leaflets to roll upward and turn yellow.

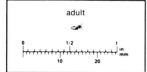

POTATO PSYLLID

adult

Potato psyllid nymphs have a flattened, scalelike appearance. When feeding, they inject a toxin that causes psyllid yellows.

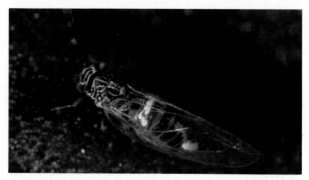

Adult potato psyllids are cicadalike in appearance. Although their feeding does not damage potato plants, their presence indicates a need to check for nymphs.

psyllids but will reduce populations of natural enemies, such as pirate bugs and damsel bugs, that occasionally feed on plant sap.

During feeding, potato psyllid nymphs inject a toxin which causes a disease known as *psyllid yellows*. Young leaves on affected plants are abnormally erect, their basal portions are cupped, and they often turn red or purple. Affected plants have shortened internodes; older leaves become abnormally thick, roll upward, and turn yellow. Plants affected early in the season may be severely stunted or may die. Diseased plants produce a large number of abnormally small tubers which sprout without a dormant period; small aerial tubers may also appear in leaf axils. In fields with a significant number of affected plants, the number of marketable tubers is greatly reduced; tubers from affected plants are not suitable for use as seed.

As few as 3 to 5 psyllid nymphs per plant can produce early symptoms, but 15 or more are needed to produce severe symptoms. Low populations are damaging before and during tuber initiation, but once tubers are formed plants tolerate injury. Feeding by adult psyllids has little or no effect at any time.

In Colorado, damaging populations occur every few years—regularly enough to necessitate monitoring every season. Psyllid adults are attracted by the same pan traps used for aphids; sweep nets are commonly used in Colorado. Presence of adults indicates a need to check for nymphs. Populations in potato fields may be monitored by taking sample leaves in the way described for aphids. Like green peach aphids, psyllid nymphs are concentrated on lower leaves.

There is no specific treatment threshold for potato psyllid. Some of the insecticides used for aphid control also kill psyllids, but special applications may be needed when psyllids appear earlier in the season than aphids. Aldicarb is not effective against this pest.

Twospotted Spider Mite
Tetranychus urticae

The twospotted spider mite occurs everywhere in the western United States, but significant damage to potatoes occurs only occasionally. The mites injure potato vines by puncturing the surface cells of leaves. Injured leaves develop small yellow splotches that darken to reddish brown and may run together to cover most of the leaf surface; entire leaves may become dried out and brittle. Injury often is not noticed until reddish brown patches of affected plants appear in the field. Injury is most common in hot, dry weather and seldom occurs before midseason. Damage is rare in cooler areas. Severe injury may lower yield by reducing the capacity of plants to perform photosynthesis.

The twospotted spider mite has two red eyespots near the head and a dark blotch on each side. The mites are tiny, and a hand lens is required to see these features. Natural enemies, such as the western predatory mite shown here attacking a spider mite, usually keep spider mite populations under control.

Leaflets injured by spider mites have yellow blotches and often become dry and brittle. A thin webbing covers heavily infested leaves.

Minute pirate bugs are predators that feed on thrips, psyllids, small aphids, and insect eggs, as well as the mite eggs shown here.

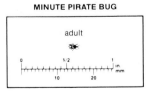

SPIDER MITES	MINUTE PIRATE BUG
adult	adult

Spider mites are so small that they appear to the naked eye only as tiny, moving dots, although you can see them easily with a 10× hand lens. Adult females, the largest forms, are about 1/60 inch (0.45 mm) long. A single leaf may support many hundreds of individuals. An adult mite has eight legs and an oval body with two red eyespots near the head. Immatures resemble adults and feed in the same way. The name spider mite comes from the loose, thin webbing the mites produce on infested leaves.

Twospotted spider mites feed on a large number of crops and weeds. In cold areas, adults overwinter in soil or in debris on the ground. Infestations in potatoes often begin with adults carried by wind from infested crops such as beans, corn, alfalfa, seed clover, or orchards. Injury often appears first at the upwind edge of the field, then spreads downwind. Dust favors the mites; infestations are often concentrated along dusty roads.

Spider mites are often kept under control by natural enemies, including insect predators such as thrips and minute pirate bugs as well as predatory mites. Secondary outbreaks of spider mites may occur when natural enemies are destroyed by insecticides applied for aphids or other insect pests.

When control is needed, a single application of an acaricide is usually sufficient. In some cases, it is necessary only to treat a strip along the edge of the field. Sprinkler irrigation helps to limit mite damage by increasing humidity in the plant canopy, making conditions less favorable for the mites.

CHEWING INSECTS THAT DAMAGE FOLIAGE

Insects such as beetles and caterpillars can reduce yield if they destroy enough foliage to reduce the ability of the plant to perform photosynthesis. Injury to leaves is most damaging during tuber initiation, when a relatively small amount of feeding may reduce the amount of carbohydrate that is available to support tuber initiation and growth. Later in the season, once a dense canopy is developed, plants have more than enough foliage to absorb all the available light. At this stage, plants must lose a significant proportion of their foliage before the damage interferes with light reception and reduces photosynthesis enough to affect yield.

Many chewing insects are controlled by insecticides applied for control of green peach aphid. Special treatments are seldom necessary under western conditions except in areas where the Colorado potato beetle reaches damaging levels.

Colorado Potato Beetle
Leptinotarsa decemlineata

The Colorado potato beetle occurs in several western states (Figure 31), and is an important pest in Idaho, Montana, Oregon, Washington, Wyoming, and northeastern Colorado. Larvae and adults feed on potato foliage; heavy infestations can damage vines severely. Although some insecticides applied for green peach aphid are also effective for potato beetle, extra control measures are needed in many cases.

Description

The oval, convex adult beetle has light yellow wing covers with 10 black stripes; the rest of the body is orange except for black markings on the head and first segment of the thorax. The yellow to orange eggs are laid on end in clusters of 10 to 30, mostly on the undersides of leaves. They resemble eggs of ladybird beetles. Young larvae are dark red with shiny black heads and legs. Older larvae are yellow, orange, or pink, with two rows of black dots along each side. The orange or red pupa is formed in a cell in the soil.

Seasonal Development

Depending on the length of the growing season, the Colorado potato beetle has one to three or more generations a year. Adults overwinter in the soil, emerging in spring at about the time volunteer potatoes begin sprouting. Early volunteers are often heavily infested, and nightshades are common early season hosts. Adults can fly several miles in search of host plants.

Once they find a host plant, the beetles feed, mate, and begin laying eggs. Each female may lay several

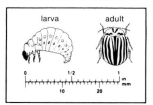

COLORADO POTATO BEETLE

larva   adult

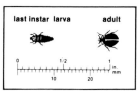

LADY BEETLE

last instar larva   adult

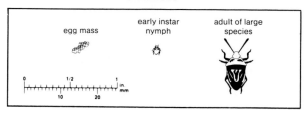

STINK BUGS

egg mass   early instar nymph   adult of large species

Colorado potato beetles are orange with black-striped, yellow wing covers and black markings on the head. The female lays clusters of yellow to orange eggs, usually on the undersides of leaflets. Ladybeetle eggs are easily confused with Colorado potato beetle eggs; however, potato beetle eggs are larger and adults usually remain nearby.

Colorado potato beetle larvae and adults chew irregularly shaped holes within and along leaf margins. When populations are high, they can completely defoliate plants.

The twospotted stink bug is usually yellow or red, with a black "Y" on its back and two black spots on its thorax. This one is attacking a Colorado potato beetle larva, but stink bugs also feed on cutworms and armyworms.

The newly hatched Colorado potato beetle larva is dark red with shiny black head and legs.

Older Colorado potato beetle larvae are yellow, orange, or pink, and have two rows of black dots along each side.

Ladybeetle adults and larvae feed on aphids as well as the eggs of Colorado potato beetle and other insects. The larva of *Hippodamia convergens* is shown here (*left*). Ladybeetle eggs are similar in appearance to those of Colorado potato beetle, but smaller (*right*). Ladybeetle eggs usually are not present before midseason.

Figure 31. The Colorado potato beetle occurs in the potato growing areas indicated in black. However, it is not a major pest in New Mexico or Utah.

hundred eggs over a period of a month or more. The eggs hatch in 4 to 10 days, depending on temperature, and the larvae feed for 2 or 3 weeks. Larvae prefer tender young buds and leaves, but feed on increasingly older leaves as they grow. At maturity, each larva burrows into the soil to pupate, forming a pupation cell 4 to 10 inches (10 to 25 cm) below the surface. The pupal stage lasts about a week.

In most areas, adults from first-generation pupae emerge in a few days and lay eggs for a second generation. In the coolest areas, however, there is only one generation a year; these adults emerge from the pupation cell, feed for 7 to 10 days, then reenter the soil, where they overwinter. In the warmest areas, there are three or more generations and the overwintering period is much shorter.

The potato beetle was first found in the mid-nineteenth century in the area of what is now the Iowa-Nebraska border, before potatoes were cultivated there. It fed on buffalobur, *Solanum rostratum*, a native plant in the potato family. Most potato family weeds, including black nightshade, cutleaf nightshade, hairy nightshade, and groundcherries, are also potato beetle hosts. Black nightshade is attractive to egg-laying females, but is toxic to the larvae. Weeds and native host plants serve as a reservoir of Colorado potato beetle; there is always at least a small population on these hosts that will move to

potato plants when they are available. Tomato, pepper, eggplant, and tobacco are potato beetle hosts. Tomatoes can be important alternate hosts in Utah; otherwise no cultivated plant is an important part of the reservoir in the western United States.

Damage

Most potato beetle injury is due to the larvae, although adults also feed on foliage and buds. Loss of leaf surface reduces the ability of plants to produce carbohydrates for storage in tubers. Heavily defoliated plants have reduced yield and lower tuber quality. The degree of loss depends on the extent of defoliation and the time of season at which it occurs. The greatest losses occur when there is prolonged, extensive defoliation during the period when tubers are forming. Before and after tuber formation, plants can tolerate moderate defoliation with little or no loss of yield.

Management

Control of Colorado potato beetle is needed most often in areas with two or more generations. In short-season areas, winter mortality often prevents the single generation from reaching damaging numbers. Also, certain systemic insecticides applied for control of green peach aphid frequently control the first generation of potato beetle.

There are no specific treatment thresholds for Colorado potato beetle. When an insecticide is needed, the timing depends on whether you are treating the first or second generation. If you are treating the first generation in an area with two or more generations, spray as soon as you find mature (fourth-instar) larvae. If you wait too long, some larvae will complete feeding and enter the soil to pupate, escaping the treatment and emerging later to begin the second generation. If you treat too early, overwintering adults may emerge after the spray residue has lost effectiveness. If you are treating the second generation, spray when all the eggs have hatched but while most larvae are still small, before they cause substantial defoliation. It is important to wait until all eggs have hatched; some defoliation at this stage is not harmful. Using a material with relatively long residual action reduces the chance that a second treatment will be needed later.

Colorado potato beetle has shown substantial resistance to major insecticides in the eastern United States (see Table 11), although resistance has not yet become a problem in the West. To reduce the chance that resistance will develop here, limit your insecticide treatments to situations where they are necessary to prevent economic loss and rotate the types of chemicals you use.

Elimination of culls and volunteer potato plants, as required for control of viruses and other diseases, helps to limit early potato beetle populations. Tillage and crop rotation reduce the number of overwintering adults emerging in the spring. Cultural practices alone do not prevent infestation, however, because the beetles are capable of flying long distances from weeds and native hosts to reach potato fields.

Natural enemies of the potato beetle include lady beetles, which feed on the eggs and young larvae, and twospotted stink bugs, which attack eggs and larvae. Lacewing larvae and collops beetles also feed on potato beetle larvae. Tachinid flies are common parasites in some areas. However, natural enemies seldom have enough impact on potato beetle populations to prevent them from reaching damaging levels.

Cutworms

Cutworms are caterpillars that are active mainly at night. During the day, they usually hide in soil cracks, under clods, or in debris on the ground. In a potato field with a good plant canopy, they hide under the plants on the soil surface. Cutworms feed mostly on foliage, but they may also chew shallow holes in exposed tubers or cut off stems of small plants near the soil surface.

The most common cutworm species that damage potatoes in the western United States are the black cutworm, *Agrotis ipsilon*, the variegated cutworm, *Peridroma saucia*, the spotted cutworm, *Amathes c-nigrum*, the army cutworm, *Euxoa auxiliaris*, and the red-backed cutworm, *E. ochrogaster*. Most cutworms are about 1 to 2 inches (2.5 to 5 cm) long when mature; they vary in color, but generally appear smooth-skinned and often have a wet or greasy appearance. Many species curl into a C when disturbed.

Cutworm adults, drab gray or brown moths that are active at night, lay their eggs on stems or on the soil surface. Larvae of most species feed mainly on plant parts in or near the soil, although some species may climb onto potato vines at night to feed on higher foliage. Pupation occurs in a cell in the soil. Cutworms overwinter in the soil either as larvae or pupae. There may be from one to three generations a year, depending on the species and the length of the season.

Cutworm damage is sometimes blamed on loopers or armyworms, which are usually present on plants in the daytime if their damage is noticeable. If you see plants with ragged holes in the leaves but you cannot find caterpillars or other chewing pests on foliage, check for cutworms in the soil or on the soil surface under the plants.

Where systemic insecticides are applied to the soil, cutworm injury seldom appears until the latter half of the season, when damage to foliage has little effect. Damage to tubers and stems is spotty. Insecticides applied to foliage for control of green peach aphid, potato beetle, and other pests usually kill cutworms also. Cutworms are commonly controlled by parasitic wasps and tachinid flies, predators, and diseases.

Loopers
Trichoplusia ni and *Autographa californica*

Loopers feed mostly on older leaves, where mature larvae may chew large, ragged holes. Damage usually occurs late in the season and it seldom affects yield even when large numbers of loopers are present. Cutworm injury is sometimes confused with looper injury since the two species

Most cutworms, such as the variegated cutworm shown here, appear smooth-skinned and curl into a C shape when disturbed.

CUTWORM

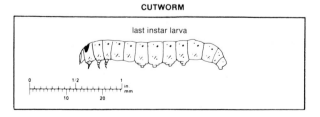

Loopers are usually light green with narrow, pale stripes along the back and sides. They arch into a loop as they crawl. Unlike cutworms, loopers are found on potato plants during the day.

The western yellowstriped armyworm has black and yellow stripes along the entire length of the body, and a velvety black spot on the side of the first abdominal segment.

Loopers chew large holes, mostly in mature leaves.

The parasitic wasp, *Hyposoter exiguae*, lays its egg in a young armyworm larva. This wasp is an important natural enemy of armyworms and loopers.

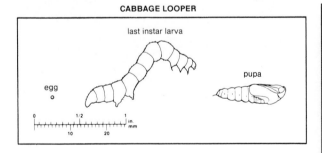

CABBAGE LOOPER

last instar larva

egg

pupa

YELLOWSTRIPED ARMYWORM

last instar larva

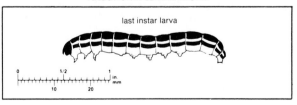

HYPOSOTER

adult

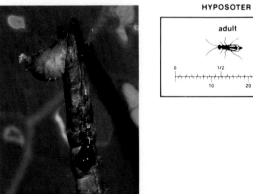

Outbreaks of nuclear polyhedrosis virus can reduce looper populations rapidly. The limp bodies of loopers killed by the virus are left hanging from foliage, often oozing their contents.

may be present at the same time, but loopers are readily found on plants in the daytime while cutworms hide under plants or in the soil.

The two looper species found on potato, cabbage looper (*Trichoplusia ni*) and alfalfa looper (*Autographa californica*), are very similar. Both are usually light green with several narrow, pale stripes along the back and sides. A small proportion of alfalfa loopers are gray instead of green, with a black head and dark stripes. Loopers are easily recognized by their habit of arching into a loop as they crawl. They have only two pairs of prolegs in the middle of the body, whereas most other caterpillars have four pairs.

In most western potato-growing areas, loopers have two or three generations a year. They feed on a variety of weeds and on alfalfa, beans, vegetables, and cotton as well as on potato. Loopers overwinter as pupae in the soil.

Natural enemies normally keep loopers from reaching damaging levels. In some areas, loopers appear only when systemic insecticides are used. Looper eggs and small larvae are attacked by bigeyed bugs, minute pirate bugs, and other predators. *Trichogramma* parasites kill the eggs, and several other parasites, especially *Hyposoter exiguae* and *Copidosoma truncatellum*, attack the larvae. Loopers are also subject to a nuclear polyhedrosis virus which can reduce populations rapidly. Common insecticides applied to foliage for green peach aphid and Colorado potato beetle also control loopers.

Yellowstriped Armyworms
Spodoptera ornithogalli and *S. praefica*

The two yellowstriped armyworm species occasionally migrate to potato from infested alfalfa, usually just after the alfalfa is cut. When numerous, the larvae can defoliate plants at the edge of a field, and they may damage exposed tubers in the same way as cutworms. To stop migrating larvae, plow a trench with the steep side toward the potato field, then apply an insecticide to kill larvae in the trench. Large-diameter irrigation pipe or a strip of stiff aluminum foil set on edge in the soil can also act as a barrier. Insecticide baits are also available. Strip or spot treatments may be needed at the edge of a field, but it is rarely necessary to treat a whole potato field for yellowstriped armyworms. Occasional fieldwide outbreaks of armyworms in California have been associated with the use of systemic insecticides.

The two species of yellowstriped armyworm are too similar to distinguish in the field; both have contrasting black and yellow stripes along the entire length of the body, and both have a velvety black spot on the side of the first abdominal segment. *Spodoptera praefica*, known

as the western yellowstriped armyworm, is more common in most potato-growing areas of the western states; *S. ornithogalli* is found in southern California, Arizona, and Utah.

Grasshoppers

Normally, grasshoppers are of little concern in cultivated crops, but in seasons when conditions favor them, large populations may build up on rangelands and grasshoppers may then migrate to fields in damaging numbers. During such outbreaks, grasshoppers can defoliate potatoes and other crops.

Common species that affect potatoes in the West include the migratory grasshopper, *Melanoplus sanguinipes*, the redlegged grasshopper, *M. femurrubrum*, the differential grasshopper, *M. differentialis*, the twostriped grasshopper, *M. bivittatus*, and the clearwinged grasshopper, *Camnula pellucida*. Other species are found locally.

Damage is usually limited to a short period in summer, since there is only one generation a year. When grasshoppers move into a field from adjacent land, you can usually control them by spraying the edge of the field or the fencerows or roadsides closest to the source of migration. Insecticidal baits are effective when applied on an areawide basis to rangelands before migration occurs.

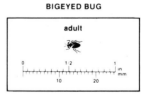

BIGEYED BUG

adult

Bigeyed bugs feed on small caterpillars, mites, aphids, leafhoppers, and on the eggs, small nymphs, and young larvae of several other insect pests.

The tip of the wireworm abdomen is flattened and has a pair of short hooks. The sugarbeet wireworm is shown here.

WIREWORM

larva

Wireworms are slender, yellowish to brown larvae, with 6 short legs that are close together near the head. This species is the sugarbeet wireworm.

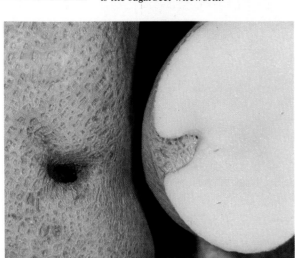

Tunnels made in tubers by wireworms are round, usually about 1/8 inch (2 to 3 mm) in diameter, and are usually lined with skin (periderm).

PESTS THAT DAMAGE TUBERS

With the exception of the potato tuberworm, insects and related pests that feed directly on tubers are soil-dwelling species; injury may occur with little or no aboveground sign of infestation. Infestations are usually spotty, but they often recur from year to year in the same places. Keep records of injury so you can plan preventive measures for subsequent crops.

Wireworms

The soil-dwelling larvae of click beetles, wireworms bore into seed pieces and tubers, causing both internal and external damage. They occur throughout the western United States, but significant damage is uncommon except in Idaho, Oregon, Washington, and Utah. Control is based on the application of insecticides or fumigants to the soil.

Description

Several wireworm species occur in western potato soils, but the most common are the sugarbeet wireworm, *Limonius californicus*, the Pacific Coast wireworm, *L. canus*, and the Great Basin wireworm, *Ctenicera pruinina*. All are similar in appearance. Wireworms resemble the mealworms sold as pet food; they are slender, elongate, yellowish to brown larvae with smooth, tough skin. The body is nearly cylindrical, but flat on the lower side. The six short legs are close together near the head, and the tip of the abdomen bears a flattened plate with a pair of short hooks. The slender, brown to black adults are known as click beetles because of their ability to flip themselves over with an audible click when placed on their backs.

Seasonal Development

Common wireworm species require 3 to 4 years to complete their life cycle. Most of the time is spent in the larval stage, but all stages may be present at once during the growing season. Larvae feed on the roots of many crops and weeds and bore into stems and other plant structures such as potato tubers. The larvae move up and down in the soil in response to temperature and moisture. In potato fields, they generally spend most of the growing season in the upper few inches of soil. If the temperature in shallow soil exceeds 80°F (27°C), however, they move deeper. In areas with cold winters, they overwinter at depths as great as 2 feet (61 cm), then return to surface layers when the soil temperature reaches about 50°F (10°C). Most damage to potato is caused by larvae in the second and third years of development.

Mature larvae form a pupation cell of soil particles; they may pupate right away or may remain in the cell over the winter, pupating in the spring. Adults that develop in the fall may also remain in pupation cells over the winter. In spring and summer, adults burrow to the surface. Both sexes are capable of flying to reach mates and egg-laying sites. Females burrow back into the soil to lay eggs.

The Pacific Coast and sugarbeet wireworms are found mostly in irrigated soils. Large numbers often build up in fields planted to cereals. The Great Basin wireworm tolerates drier, colder soil conditions than the others; it is the most damaging species in newly cultivated desert soils and in fields recently converted from dryland farming or pasture to row crops such as potato. Infestations of Great Basin wireworms decline over 2 to 4 years of irrigation as the initial population matures and emerges as adults. They may be replaced, however, by sugarbeet wireworms or other species that favor irrigated soils.

Damage

In spring, wireworms may bore into seed pieces and developing shoots. These injuries often become infected with fungi or other decay organisms, resulting in weak plants and spotty stands. In summer, wireworms bore into developing tubers. Entry holes are round, usually about 1/8 inch (2 to 3 mm) in diameter. Damaged tubers often look like they have been punctured with a nail. Tunnels in the tubers are usually straight and may be shallow or quite deep. They are usually lined with periderm and are seldom infected with fungi.

Management

Because wireworms have a long life cycle and feed on a variety of hosts, knowledge of the cropping history in a particular field is essential. Wireworms are often present following cereals, even though injury was not apparent in the grain crop, and they may also be present in newly cultivated soils. Although alfalfa is not a favorable host for wireworms, populations often build up in alfalfa fields that are allowed to become weedy.

Where row crops such as potatoes, corn, or beans are planted for several years, crop injury may recur each season in places where wireworms are concentrated. In these situations, it is a good idea to check for wireworms before planting potatoes.

If the soil is not too wet or too fine in texture, you can sample for wireworms with a set of two screens—the upper of 1/4-inch (6-mm) hardware cloth, and the lower of window screen. Collect samples of soil at random in as many places as possible throughout the field, sift them through the screens, and record the number of wireworms. Results are expressed in terms of the number per square foot, so you must know the area covered by each sample. For example, a typical two-handled posthole digger takes about 1/4 square foot each time. If you are sampling in the spring, make sure the soil temperature is at least 45°F (7°C) and take samples to a depth of 6 inches (15 cm). In late summer, when surface soil is dry, sample to a depth of 18 inches (45 cm).

You can also check for wireworms by baiting. Bury pieces of carrot about 3 inches (8 cm) deep at a series of random locations in the field; mark each place clearly. Check the bait in 2 or 3 days to see if wireworms are present. Another suitable bait is 2 to 3 tablespoons of coarse wholewheat flour wrapped in a small square of netting. Baiting is not effective if the soil contains a large amount of plant residue, or the soil is too cold, wet, or dry for wireworms to be active near the surface.

Insecticides for wireworm control may be applied before planting as either broadcast or banded treatments. After planting, side-dressing can be used soon after shoots begin to emerge. If the wireworm population is high, you may need a combination of a broadcast application with a band treatment or side-dressing. For broadcast treatment, apply granules or emulsifiable concentrate uniformly over the entire field area, and mix the material into the soil immediately to the depth specified on the label. For in-furrow application, place a narrow band of granules 3 to 4 inches (7 to 10 cm) below the seed pieces at planting. If you use two treatments, apply a different insecticide each time.

Treatment thresholds for preplant applications are usually based on the number of wireworms per square foot in soil samples. A threshold suggested in Washington is 0.4 wireworms per square foot. There are no thresholds based on baiting.

To determine if a side-dressing is needed after planting, walk the field as soon as shoots begin to emerge, and dig up seed pieces at random throughout the field to check for wireworm injury. This is a good way to check the effectiveness of preplant treatments. There is no specific treatment threshold for this method, however.

More information on wireworm monitoring and control is presented in *Management of Potato Insects in the Western States*, listed in the References.

Once wireworm populations have been reduced, they usually remain low unless favorable crops such as grains are replanted. Where wireworms have been a serious problem, it may be best to avoid grains in rotations with potato. Alfalfa is a good rotation crop for infested soils as long as the stand is strong and does not become weedy.

Flea Beetles
Epitrix spp.

The western potato flea beetle, *Epitrix subcrinita*, and the tobacco flea beetle, *E. hirtipennis*, are widespread in the western U. S., while the tuber flea beetle, *E. tuberis*, is found in Washington and Oregon. The three species are similar in appearance and life history. Adults are about 1/16 inch (1.5 mm) long, shiny green to brown or black, with enlarged hind legs that enable them to jump several inches when disturbed. The larvae are whitish, soft-bodied grubs with a yellowish or light brown head and six short legs; they are up to 1/4 inch (6 mm) long.

Adult flea beetles overwinter in weeds or debris outside the field. In spring, they feed on weeds until potato plants emerge, then fly into potato fields and feed on foliage. The adults chew small, round holes in leaves, giving them a shothole appearance; this injury is seldom important, although heavy infestations of tuber flea beetle can cause significant defoliation of young plants.

Most damage is caused by the larvae, which hatch from eggs scattered by adult females in the soil around potato plants. The larvae feed on roots, underground stems, and tubers, and then pupate in the soil. Development takes about a month and there are one or two generations a year.

Larvae feeding on tubers produce narrow, straight tunnels about 1/32 inch (0.8 mm) in diameter in the flesh of tubers. Fungi often grow in the tunnels, turning them black or brown. Tunnels of the tuber flea beetle may extend 1/2 inch (12 mm) into the tuber; when extensive, they make tubers unsuitable for processing. Injury caused by the other species seldom extends more than 1/4 inch (6 mm) into the tuber—shallow enough that it can be removed by peeling. Larvae of all species may also cause small, raised bumps and shallow, winding trails on the surface of tubers.

Systemic insecticides and foliar sprays applied for green peach aphid and Colorado potato beetle usually keep flea beetles below economically damaging levels. Even in areas where these treatments are not used, flea beetle infestations are sporadic and special controls are rarely necessary. Flea beetle damage often is not noticed until harvest, when it is too late for control measures.

White Grubs

The larvae of scarab or June beetles, white grubs are widespread in the western states, but significant damage to potatoes is uncommon. Species that feed on potatoes include the carrot beetle, *Bothynus gibbosus*, and the ten-lined June beetle, *Polyphylla decemlineata*.

White grubs are usually curled into a C. Most of the body consists of the whitish, translucent abdomen, and a dull gray portion is at the tip. The head and the prominent legs are brown. Larvae may be up to 1 1/4 inches (3 cm) long when mature. The carrot beetle has one generation a year; the tenlined June beetle requires 2 to 3 years to develop. Most of this time is spent as larvae. Both species pupate in the soil; adults emerge to mate and lay eggs.

White grubs chew large, shallow gouges in tubers. In some cases, these injuries may cover much of the surface. Infestations are most common in sandy soil where sod or other organic matter has been plowed in before planting. Special management practices are seldom needed in potatoes.

Leatherjacket
Tipula dorsimaculata

The larvae of certain crane flies, leatherjackets, are wormlike, legless maggots with tough, folded skin resembling leather. Mature larvae are gray to brown and up to 1 1/2 inches (3.3 cm) long. The small, pointed head retracts deeply into the thorax and may not be visible when specimens are handled; the tip of the abdomen bears a group of fingerlike projections. Leatherjackets feed largely on decaying plant debris but will also feed on potato tubers, producing round holes up to 1 inch (2.5 cm) deep. Infestations are rare except in fields following an alfalfa crop plowed under in the spring; the leatherjackets feed at first on the decaying alfalfa, moving to tubers later. Special management practices are not necessary for leatherjackets.

Seedcorn Maggot
Delia platura

The seedcorn maggot is a widespread pest that occasionally damages potato seed pieces in cool, wet soil. The maggots hatch from eggs laid in moist soil by the adult female, a drab, gray fly about the size of a housefly. The legless, whitish maggots, about 1/4 inch (6 mm) long when mature, burrow into seed pieces and underground stems; they feed for 2 to 3 weeks before pupating in the soil and emerging as adults. There may be two or three generations a year, but only the first affects potatoes.

Damage is more likely in soil where a lot of plant debris is plowed in before planting. A delay in planting can reduce the chance of injury in such fields by allowing the soil to warm up, reducing the period when seed pieces are vulnerable. Seed pieces that are well suberized are seldom damaged.

Garden Symphylan
Scutigerella immaculata

Garden symphylans are slender, whitish animals related to centipedes. They are found throughout the western United States, but infestations are usually limited to small areas with soils high in organic matter, where they persist from year to year.

Adults are up to about 3/8 inch (1 cm) long, with 10 to 12 pairs of legs and a pair of antennae. They run rapidly to escape light when exposed in soil. Eggs are laid in soil, and immatures develop to maturity through a series of molts; immatures resemble adults but have only six to 10 pairs of legs. Under favorable conditions, adults live for several years. Symphylans move up and down in the soil in response to temperature and moisture. They are most active at 50° to 70°F (10° to 21°C).

Symphylans feed on roots and other underground structures of many crops and weeds. Damage to roots can stunt potato plants. An injured tuber has small, round entry holes leading to irregular chambers beneath the skin; a ring of dark, corky tissue develops around each entry hole. Symphylans do not produce narrow tunnels such as those caused by flea beetle larvae.

Where treatment is necessary, apply an insecticide specifically recommended for symphylan control to the soil in the same way as described for wireworms. For best results, use a broadcast application as shortly before planting as possible; mix the insecticide into the soil thoroughly. Soil fumigation can protect potatoes for up to 2 years if the treatment is thorough. Continuous flooding for 3 weeks or more in summer will also reduce symphylan infestations.

Potato Tuberworm
Phthorimaea operculella

The potato tuberworm is a worldwide pest of stored potatoes, but it infests potatoes outdoors only in warm climates. In the western United States, field infestations occur regularly only in California from Stockton and Salinas southward (Figure 32), although the pest has been found in Arizona, New Mexico, and southwestern Utah. Outdoor infestations elsewhere in the western region usually originate in infested tuber storage. In California, cultural practices can greatly reduce field infestations; insecticide treatments can be scheduled with pheromone traps.

Description

The tuberworm is a small caterpillar, the larva of an inconspicuous gray and black moth called the potato

OSU EXTENSION SERVICE

Larvae of the tuber flea beetle make narrow tunnels in tubers. Some of the tunnels turn brown or black in response to fungal invasion.

Garden symphylans are white and run rapidly to escape light when uncovered.

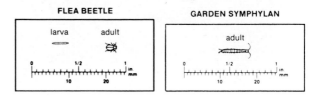

FLEA BEETLE

larva   adult

GARDEN SYMPHYLAN

adult

The potato tuberworm is a small, light-colored caterpillar with a brown head. Newly hatched tuberworm larvae create small, hollowed out blotches in the surface of potato leaves.

The inconspicuous, gray tubermoth is active at dusk or during the night. It hides during the day in sheltered parts of the plant or on the ground.

Leaflets damaged by tuberworms are usually curled and shriveled.

Tuberworm larvae feed just below the skin of the tuber at first but eventually tunnel deep into the flesh. Tunnels turn black with larval feces and fungal growth.

POTATO TUBERWORM

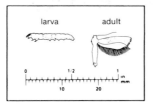

larva   adult

0   1/2   1 in
                            mm
   10   20

Water pan traps for monitoring tuberworm populations use pheromone lures to attract adult males.

tubermoth. A larva is usually dull white with a brown head; a mature larva may have a pink or greenish tinge. The smooth, brown pupa is formed in a silk cocoon covered with soil particles or bits of plant debris. The spherical, whitish to brown eggs are seldom noticed because of their small size.

Seasonal Development

The tuberworm has several generations a year in California; development continues all year as long as host material is available. In addition to potato, tuberworms feed on potato family plants including tomato, eggplant, black nightshade, silverleaf nightshade, and jimsonweed. Most infestations, however, originate in infested potato plants or tubers in potato fields and cull piles.

Adult females lay eggs on foliage, soil, plant debris, or exposed tubers; the moths will crawl through soil cracks or burrow a short distance through loose soil to reach tubers. On tubers, young larvae feed just below the skin at first, but eventually tunnel deep into the flesh. A newly hatched larva on foliage usually begins feeding between the surfaces of a leaf, creating a small, hollowed-out blotch; later, larvae sometimes fold sections of leaf into shelters fastened together with silk. Larvae may also bore into petioles or stems. Generally, however, there is relatively little foliage feeding by tuberworms. Larvae often move from foliage to tubers, leaving the first feeding site and crawling or dropping down to the soil; they do not reach tubers by boring through stems or roots.

When they have finished feeding, larvae spin silk cocoons on the soil surface or in debris under the plant; pupation normally does not occur in tubers. In storage, larvae may crawl a considerable distance before pupating in crevices among building materials, in potato sacks, or at a similarly protected site.

Adults mate and females begin laying eggs soon after they emerge from the pupa. Females release a chemical scent, or pheromone, that attracts males for mating. Adults are active at night and at dusk. During the day, they hide in sheltered parts of the plant or on the ground.

The tuberworm completes a generation in 3 to 4 weeks in summer; about half of this time is spent in the larval stage. In winter, larvae may require 3 months or more to complete development. The developmental threshold is 52°F (11°C); about 700 degree-days are required for a generation. Larvae and adults can survive long periods at temperatures near freezing; feeding and reproduction resume whenever the temperature again reaches 52°F.

Damage

Tuberworms create deep tunnels about 1/8 inch (3 mm) in diameter through the flesh of tubers. The tunnels

Figure 32. The potato tuberworm is an important pest in the growing areas indicated in black.

usually appear black, because they are filled with larval feces and are often infected with fungi. Infested tubers are unmarketable. Injury to foliage is rarely extensive enough under California conditions to be damaging.

Management

Anything that reduces exposure of tubers to egg-laying females will reduce tuberworm damage. The moths generally cannot reach tubers covered with 2 inches (5 cm) or more of soil unless the soil has deep cracks.

Potato varieties that set tubers on relatively deep stolons are less vulnerable to infestation; shallow-setting varieties such as Kennebec are more vulnerable. A low level of injury is more important in tubers intended for the fresh market than in those intended for processing.

Sprinkler irrigation is valuable in keeping the soil surface sealed, especially in fine-textured soils that tend to develop deep cracks as they dry. Use sprinklers to keep the upper layer of soil moist enough so it will not crack. Tuberworm damage is almost always more severe in furrow-irrigated fields. Where it is necessary to use furrow rather than sprinkler irrigation, make sure the field is properly graded and keep furrows in good shape to prevent washouts that expose tubers.

Prompt, thorough harvest and sanitation are also essential. Harvest tubers as soon as possible after they have matured; do not leave them in the ground longer than necessary. In the Salinas Valley, where tubers are

DATE	MTN THIS CHECK	CUMULA-TIVE MTN	NUMBER OF CHECK DATES	SEASONAL AVERAGE
MAR 1	2	2	1	2
4	5	7	2	3.5
8	3	10	3	3.3
11	6.2	16.2	4	4.1
15	7	23.2	5	4.6
18	9.4	32.6	6	5.4
22	8	40.6	7	5.8
25	10.6	51.2	8	6.4
29	10.8	62.0	9	6.9
APR 2	14	76.0	10	7.6
5	16	92.0	11	8.4

Figure 33. In monitoring potato tuberworm, keep a chart such as this for each field to record the results of pheromone trapping. On each check date, record the number of moths per trap per night (MTN) caught since the previous check. Divide the cumulative total of the MTN values for the season to date by the number of check dates to find the seasonal average MTN.

stored in the ground, use additional hilling after vine killing to keep tubers covered with at least 2 inches (5 cm) of soil. During harvest, avoid leaving tubers on the surface overnight. After harvest, make sure that all unharvested or discarded tubers are deeply buried or destroyed; never leave cull piles in or near the field or around processing or storage facilities. Eliminate volunteer potato plants from fields, waste areas, and from stands of other crops following potato.

Monitoring and Control. Commercial pheromone lures for potato tuberworm moths can be used to monitor the pest either in water-pan traps or delta traps. Water pan traps consist of a pan about 8 inches (20 cm) in diameter, filled with water to a depth of at least 3 inches (7.5 cm); the lure is placed beneath a piece of sheet metal that is bent to form a tent-shaped cover over the pan. Male moths attracted to the lure fall into the water; the water must contain a few drops of soap to break the surface tension and prevent moths from escaping. A delta trap is a three-sided cardboard shelter with a sticky inner surface to catch the moths. Delta traps must be fastened to a stake set in the ground.

Pan traps are generally best for tuberworm monitoring, as they are easily cleaned and refilled at each check,

and they are not affected by sprinkler irrigation. The sticky surface of delta traps quickly becomes clogged with dust and moth scales, so the traps must be replaced frequently; when they are dusty, delta traps do not retain the moths attracted to the lure.

Treatment Thresholds Using Water Pan Traps. Use at least four pan traps per field, placing one or more in each quarter of the field. Put the traps between plants on the tops of beds, and keep them at least 14 rows or 50 feet (15 m) in from the edge of the field. Check the traps every 3 or 4 days to count the trapped moths and replenish water in the traps. In hot weather, be sure to check often enough to keep the water from evaporating between checks. For each field, record the number of moths per trap per night (MTN), and keep the results in a chart (Figure 33).

The treatment threshold varies according to potato variety and field conditions. Fields where tubers are well protected by soil can tolerate more moths than fields where many tubers are exposed. In fresh market potatoes, treat whenever the MTN for a single check period reaches 15 to 20 *or* when the average MTN for the season exceeds 8. For Kennebec chipping potatoes, the threshold for one check period is 40 MTN; the seasonal average MTN should remain below 15 to 20.

If moth activity does not reach the threshold level before vinekill, do not treat. Insecticides applied at vinekill do not reduce tuberworm damage.

In addition to pheromone trapping, take a random sample of harvested tubers from each field to determine the proportion damaged by tuberworm. Check at least 100 tubers taken at random from each quarter of the field. Pay close attention to green tubers; they were exposed and will show more damage. Keep this information as a guide to treatment thresholds the following year: if there is too much damage, you may need to treat at a lower MTN level or take extra precautions to protect tubers from exposure to the moths.

Other Insects

Various insects that feed occasionally on potato plants have little or no effect on the crop, but you should be able to recognize them so that you do not confuse them with damaging pests.

Blister Beetles

Potatoes in fields close to desert or rangeland, are occasional sites for blister beetle infestation. The larvae develop as parasites on the eggs of grasshoppers or ground-nesting bees. The adult beetles cluster near the tops of plants, usually in small areas of the field; they feed on leaves and flowers. Treatment is rarely if ever necessary.

Blister beetles are easily recognized by the narrow, necklike first segment of the thorax. The species found on potato in the western states are the spotted blister beetle, *Epicauta maculata*, the ashgray blister beetle, *E. fabricii*, the punctured blister beetle, *E. puncticollis*, and the Nuttall blister beetle, *Lytta nuttalli*.

Lygus Bugs *Lygus elisus* and *L. hesperus*

Lygus populations often develop on alfalfa or weeds. When the alfalfa is harvested or the weeds dry out, adult lygus bugs may migrate to crops, including potatoes. Feeding by lygus bugs may damage young buds and sometimes causes distorted growth or wilted leaves, but generally does not affect yield. Often, the damage is not noticed until the bugs have already left the field. Lygus bugs sometimes can be an early season problem on potatoes in the San Luis Valley in Colorado.

Intermountain Potato Leafhopper
Empoasca filamenta

The intermountain potato leafhopper is the most common of the various leafhoppers found on potatoes in the western states. Adults fly into potato fields after the population has passed through one or more spring generations on weeds. The leafhoppers feed mostly on the undersides of lower leaves, causing small chlorotic spots that give leaves a speckled or stippled appearance. Leafhoppers occasionally cause problems in potatoes by transmitting curly top virus or mycoplasmas causing aster yellows or witches' broom. Insecticides applied for aphids and other pests normally keep leafhopper numbers low in potato.

Thrips

Thrips are tiny, slender insects that feed by rasping the surface of leaves. The species found on Western potatoes include the western flower thrips, *Frankliniella occidentalis*, and the onion thrips, *Thrips tabaci*. Their feeding causes brown scarring on the undersides of leaves; severely damaged leaves dry out and drop. The damage often is not noticed until the thrips population has declined or disappeared. Thrips damage is similar to that caused by windburn or blown sand, but you can recognize it by the numerous black flecks scattered over the discolored area; these are the feces of the thrips. Thrips usually build up in weedy areas or unirrigated pastures, moving to potatoes only when other hosts begin to dry out; damage is usually limited to the outer few rows adjacent to these sources.

DISEASES

Potatoes are susceptible to a large number of diseases. Several diseases occur in all growing areas; some are limited by climate or other factors to one or a few growing regions. Usually, a few diseases are most important in each area. Before you plant, find out which diseases are likely to be problems. Whenever possible, review records of disease incidence in potatoes and other crops that have been grown in the field you intend to plant. Choose varieties and plan cultural practices that will reduce the impact of key diseases. Use certified seed tubers and follow good seed-handling practices; this helps reduce many of the diseases that are problems in potatoes.

Diseases can be classified as biotic or abiotic depending on the nature of the causal agent. *Biotic diseases* are caused by pathogens—fungi, bacteria, mycoplasmas, viruses, or parasitic plants such as dodder—that invade host plants and disrupt their normal functions. *Abiotic diseases* (physiological disorders) are caused by physical factors—environmental stresses, toxic substances, or nutrient deficiencies. Physiological disorders are discussed in the next chapter.

To effectively manage biotic diseases, you must know where pathogens originate (primary inoculum sources), how they disperse, how they infect potato plants, and what environmental conditions favor disease development. Pathogens may be classified by the part of the host plant they affect and their method of dispersal. Root, stem, and tuber pathogens usually are soil-inhabiting or soil-invading organisms that are moved in contaminated soil or infected tubers; they usually infect plant parts in contact with soil. Foliar pathogens are dispersed by wind or water and infect aboveground plant parts. Viruses and mycoplasmas are spread mechanically, by insect or nematode vectors, or in tubers. Because they are carried in the vascular system, they infect all parts of the plant.

Tuber diseases include those, such as pink rot and scab, that are limited primarily to tubers. Several diseases that affect other parts of the plant also cause serious tuber diseases. These include blackleg, ring rot, potato leafroll, Rhizoctonia, early blight, late blight, and Sclerotium stem rot. In most cases, management practices provide effective control of tuber diseases. By using certified seed, suberizing seed tubers properly, using seed treatments, providing adequate but not excessive soil moisture throughout the growing season, allowing tubers to mature before harvest, handling carefully and providing correct conditions for curing and storage of tubers, you can help reduce losses to tuber diseases. Use of resistant or tolerant varieties helps reduce losses to scab and net necrosis caused by potato leafroll virus. Use of certified seed and strict sanitation practices are required to control ring rot.

Field Monitoring and Diagnosis

Before the season begins, familiarize yourself with the symptoms of potato diseases common in your area. Study the sections in this manual on specific diseases with their accompanying illustrations; they will help you know what to look for in the field.

During the season check fields regularly for disease symptoms and signs of stress. Check developing tubers for abnormal growth, damage, or decay. If symptoms of disease or stress appear, note their distribution within the field. Do they appear only on a few scattered plants? Are they concentrated in certain areas? Or are they generally distributed throughout the field? Keep records of these observations, weather, soil conditions, and previous disease outbreaks to help in diagnosis and in future planning. Soil moisture is particularly important because of its influence on tuber development and many potato diseases.

Every plant disease involves a complex interaction between the causal agent, the host plant, and the environment. Disease symptoms produced and the rate at which they develop are influenced by the genetic characteristics of both the plant and the pathogen, by the stage of growth when infection or stress occurs, and by environmental conditions, particularly temperature, soil moisture, and humidity within the plant canopy. When comparing field symptoms with illustrations and descriptions in this manual, consider the possible influence of environment or other factors on symptom expression. Examine as many affected plants as possible and look for different stages of disease to see how symptoms change during disease development. Look at all parts of the plant, including stems, roots, and tubers. Do not rely on a single symptom such as wilting or leaf yellowing to identify a disease; different diseases may produce similar symptoms if the causal agents affect the same plant function. Observation of several different symptoms is usually needed to identify a disease accurately.

Field symptoms are not always sufficient to make an accurate disease identification. When you are not sure of an identification, seek confirmation from qualified professionals. Some pathogens such as viruses may require special laboratory techniques for identification; diagnosis of nutrient deficiencies may require plant tissue analysis. However, even when laboratory analysis is required, an accurate set of field notes will help confirm the results.

Prevention and Management

Seed Selection. Use of certified seed tubers reduces or eliminates losses caused by many diseases that are tuber-borne, including potato viruses, bacterial ring rot, and several other pests. Seed tubers grown under a state's seed certification program are certified to have shown no more than certain low percentages, called tolerances, of pest and disorder symptoms during required inspections. Pests for which tolerances are enforced and the range of those tolerances among the western states are listed in Table 4 of the chapter, *Managing Pests in Potatoes*. Familiarize yourself with the certification requirements that apply to the seed tubers you buy, and obtain your seed from a source with a reputation for supplying good quality seed tubers. If possible, visit the areas where the seed tubers are grown to investigate conditions and practices that influence seed quality. Follow careful handling and sanitation procedures to reduce contamination of your certified seed tubers.

Cultivar Selection. Using resistant or tolerant cultivars is often the most convenient and effective way to reduce losses from certain diseases. Several different potato diseases and physiological disorders can be reduced by using resistant cultivars. A list of disease resistances available in potato cultivars commonly grown in the western states is given in Table 6.

Field Selection. Several potato pathogens can remain dormant in the soil for long periods and cause disease when potatoes are planted again. Some potato pathogens can cause diseases on other crops. Keep records of diseases that occur both in potatoes and in rotation crops as a guide for future management decisions. Use these records to help make decisions on what varieties to plant or what special management practices may be required. For example, in a field with a history of Verticillium wilt, you may want to avoid planting a highly susceptible cultivar such as Norgold, you may decide to fumigate, or you may choose to avoid planting potatoes. Histories of powdery mildew or severe early blight indicate a need for special fungicide applications or monitoring during the growing season. You may want to avoid planting potatoes in fields with a history of potato rot nematode or deep-pitted scab; if growing seed potatoes, avoid fields that have had problems with corky ringspot or root-knot nematodes.

Avoid fields where poor drainage or other conditions will interfere with uniform irrigation. Anything that makes water management difficult will increase the risk of damage by several potato pathogens and disorders.

Cultural Practices and Sanitation. Correct storage, handling, and planting of seed tubers, and proper management of soil moisture and fertility minimize losses to most potato diseases. Sanitation is essential to the prevention of seed piece infection during cutting and handling, and prevention of the spread of pathogens in contaminated soil, water, and field equipment. Strict sanitation requirements must be followed in growing seed potatoes. A list of cultural practices important for controlling diseases and other pests is presented in Table 3. Cultural practices and sanitation procedures are discussed in the sections on individual diseases and in the chapter, *Managing Pests in Potatoes.*

Pesticides. Fungicides can reduce damage caused by certain foliar pathogens such as powdery mildew, late blight, and severe early blight. To be effective, they usually must be applied before infection occurs or when the disease just begins to develop. Therefore, it is important to know the history of disease problems in the field, to take into account weather conditions that favor disease, and to monitor for the appearance of disease symptoms. Soil fumigants may be used to control nematodes or Verticillium wilt. Follow the guidelines given in the discussions of individual diseases and consult local authorities for current information on materials recommended to manage disease problems in your area.

ROOT, STEM, AND TUBER DISEASES

Diseases of roots, stems, and tubers are caused by fungi or bacteria that are transmitted in infected seed tubers or infested soil. Management of root, stem, and tuber diseases generally requires use of certified seed to reduce spread of tuberborne pathogens, proper suberization of seed tubers to prevent infection, careful irrigation scheduling to avoid excessively wet or dry conditions during critical growth phases, and maintenance of healthy plant growth. Crop rotation is useful for control

of soil-inhabiting pathogens that have limited host ranges and require host plant residues for survival. Rotation is less effective for pathogens such as *Verticillium* or *Phytophthora erythroseptica*, which can survive in the soil for a long time in the absence of a host. Resistant or tolerant varieties can help reduce losses caused by some soilborne pathogens.

Verticillium Wilt
Verticillium dahliae

Verticillium wilt is one of the most serious potato diseases, and commonly occurs wherever potatoes are grown in the western United States. Symptoms usually appear in mid to late season, but develop sooner and are more severe when plants are stressed by heat, low soil moisture, poor fertility, or other factors. The term *early dying* is frequently used to describe some symptoms of Verticillium wilt, because affected plants appear to be aging or dying prematurely. Verticillium wilt reduces total yields; the sooner or more severely plants are affected, the greater the yield reduction. In most growing areas, maintaining plant vigor by using proper cultural practices controls the expression of wilt symptoms and reduces losses. Soil fumigation may be used where the disease is severe.

Symptoms and Damage

Symptoms generally first appear on lower leaves. Rarely, symptoms first appear on upper leaves; this is most likely to occur when wilt develops early in the season. Areas between the leaf veins turn yellow and later brown. Affected plants may wilt on hot, sunny days and recover in the evening. Early in disease development, wilting and yellowing may affect leaflets on one side of a petiole or leaves on one side of a stem. Unilateral wilting is a short-lived symptom that usually occurs on one stem of a plant. Wilting can be caused by other diseases, particularly blackleg or ring rot; in these cases other symptoms that distinguish them from Verticillium wilt usually are present. Foliar symptoms move up the plant until entire stems turn brown and die. As Verticillium wilt progresses, diseased plants may exhibit *flagging*, standing more upright than unaffected plants.

After foliar symptoms appear, but before they become severe, the water-conducting tissue of the stem (xylem), turns a reddish to dark brown. This symptom can be seen by slicing through the stems at an angle near the soil line. Vascular discoloration is also caused by blackleg, but the color tends to be grayish brown to black and may not be restricted to xylem. Stems affected by blackleg (bacterial stem rot) are softened and usually turn black, symptoms not caused by *Verticillium*. Some xylem discoloration may be seen in plants wilting from ring rot.

Vascular discoloration caused by *Verticillium* infection may extend into the stem end of tubers. However, *Verticillium* may infect tubers without causing symptoms. Potato leafroll virus or physiological stress may cause similar tuber symptoms.

The fungus interferes with transport of water and nutrients in the xylem, so injury is most severe on coarse-textured soils and during periods of hot weather, when plants are more easily stressed for nutrients or water. Under cool conditions and when soil moisture and fertility are kept adequate, no symptoms or injury may be evident. Certain species of root lesion nematode, *Pratylenchus penetrans* and *P. thornei*, can increase the severity of Verticillium wilt. Because the disease is more pronounced after plants mature, Verticillium wilt is often more severe in areas where the season is long, such as the Columbia Basin.

Seasonal Development

Verticillium dahliae is a soil-inhabiting fungus that has many different strains causing wilt diseases in tomatoes, peppers, eggplants, melons, cotton, stone fruits, and many other crops and wild plants. In other parts of the United States and elsewhere in the world, *V. albo-atrum* causes wilt of potatoes and other crops. In the absence of a susceptible host, *V. dahliae* can survive in the soil for 8 years or more as tiny, black structures called microsclerotia. Microsclerotia are formed in dead or dying stems of infected plants and are the source of inoculum for future infections in the field.

In the presence of a susceptible host plant, the microsclerotia germinate. The fungus invades the roots through root hairs, primarily just behind the root tip, and establishes itself in the xylem, where it spreads upward into stems, petioles, and leaflets. As it grows through the vascular system, the fungus impedes water movement and causes browning of the tissues. Growth of *Verticillium* in the plant is favored by cooler temperatures of 55° to 75°F (13° to 24°C). Foliar symptoms may appear once *Verticillium* is established in a plant, although the fungus is frequently present in healthy-looking potato plants without causing symptoms. Periods of high temperatures will bring on wilt symptoms, because the disease interferes with water movement in the xylem. Damage usually occurs when plants are stressed by insufficient irrigation, deficient nitrogen, or other diseases.

The fungus is present in most soils where at least 2 or 3 seasons of potatoes have been grown under short rotations. Inoculum in the soil increases each time a crop of infected potato vines is plowed under. *Verticillium* moves from place to place in irrigation water, in wind-blown soil, on contaminated equipment, and in infected seed potatoes. Infected seed potatoes are thought to be the primary source of *Verticillium* in new ground.

Leaves of plants affected by Verticillium wilt turn yellow, then turn brown as leaf tissue dies. As plants become severely affected, they tend to stand up higher than other plants in the field, a symptom called flagging.

The water-conducting tissue (xylem) of stems infected by *Verticillium* turns reddish or dark brown. To see this symptom, slice through stems at an angle near the soil line.

Management Guidelines

Resistance to Verticillium wilt varies among potato cultivars. Nooksack and Targhee are resistant; Butte, Centennial, Kennebec, Russet Burbank, and White Rose are moderately susceptible; Norgold and Norland are susceptible. Most cultivars express a tolerance to this disease when proper soil fertility and moisture are maintained. When Verticillium inoculum in the soil is high and conditions are favorable for expression, even the most resistant commercial cultivars will develop symptoms of Verticillium wilt.

Maintaining plant vigor with good fertilization and irrigation practices is the best management presently available for Verticillium wilt. Proper soil fertility suppresses development of the disease and reduces losses; adequate nitrogen and potassium are particularly important. Irrigating with sprinklers rather than furrows makes it easier to apply precise amounts of nitrogen and water. In some cases, all plants in a field may be infected with Verticillium without showing disease symptoms; unless they are stressed, they may not be damaged.

Because Verticillium microsclerotia survive in the soil for 8 to 14 years, rotations of at least 5 years are needed to reduce the amount of Verticillium inoculum in the soil. However, rotating into field corn or legume crops other than red clover for 2 or 3 years does keep the level of Verticillium in the soil from increasing. Recommended rotation crops include corn, sugar beets, onions, alfalfa, and peas; late plantings of Sudangrass following grain or peas are also useful. Although Verticillium may survive on the roots of grain crops, grain rotations are better than no rotations at all. Ask your local extension agent, farm advisor, or other experts for recommended rotation crops in your area. Control weeds and volunteer potatoes in rotation crops, because they may be hosts for Verticillium.

Soil fumigation may be cost effective for coarse-textured soils where wilt is serious. However, Verticillium can build back to damaging levels after one season of potatoes following a soil fumigation. Fumigation is more efficient in coarse-textured soils; it is usually not economical for medium or fine-textured soils. If you are not sure fumigation is necessary for a particular field, fumigate a test strip to see how it affects yields. Certain species of root lesion nematode may increase the severity of Verticillium wilt. In some areas, nematicide treatments will reduce the amount of wilt.

Sanitation of field equipment to avoid introduction of contaminated soil and burning or removal of vines before harvest are especially important in fields that are planted to potatoes for the first time. Complete burning of vines is a very effective way to reduce Verticillium inoculum in the soil, because infected stems are the main source of microsclerotia. However, burning is not used routinely, primarily because of cost.

Blackleg
Erwinia carotovora

Known also as *bacterial stem rot*, blackleg occurs wherever potatoes are grown. The severity of the disease depends on seed-handling techniques, soil moisture and temperature at planting, environmental conditions, cultivar, amount of infection in the seed lot used, and external sources of the bacteria such as irrigation water and cull piles. Blackleg can be a major disease problem in most areas if more susceptible varieties such as Norgold and Kennebec are grown. The disease is best controlled by warming seed tubers before cutting, and planting when soil is at least 50°F (10°C) and moist but not wet. Careful selection of seed stocks and good sanitation during seed cutting help reduce the incidence of blackleg.

Symptoms and Damage

Symptoms of blackleg may appear at any time during the season. On older plants, yellowing between leaf veins and browning and upward curling of leaf margins are the first symptoms. When young plants are affected, stunting and erratic growth may occur. If the soil is excessively wet or cold and dry after planting, the seed piece will decay entirely before emergence, resulting in reduced stands. Under some conditions plants may appear stiff and erect. Occasionally aerial tubers form. Leaf curling, yellowing, and aerial tubers may also be caused by Rhizoctonia, aster yellows, or waterlogging.

Infected plants may wilt during hot weather. Entire stems may wilt without discoloration of leaves or wilting may occur with yellowing of leaves, similar to Verticillium wilt. Slicing the stems should show a grayish brown to black discoloration that is not confined to vascular tissue but sometimes spreads into the pith. As this discoloration progresses, stems may turn an inky black and become soft and mushy just above and below the soil line. This inky black symptom, which gives the disease the name *blackleg*, usually occurs during the course of the disease, but sometimes stems will become soft and mushy without the black discoloration.

In some cultivars, wilting and leaf symptoms caused by the blackleg bacterium may appear similar to symptoms caused by ring rot, but tubers of plants affected by blackleg do not show the vascular decay that may be caused by ring rot bacteria.

Decay of tubers is caused by the same bacteria that cause stem symptoms; tuber decay is not always associated with plants that show blackleg symptoms. Tuber symptoms range from a slight vascular discoloration to complete soft rot. Lesions usually begin as light-colored, odorless, soft decay, sharply separated from undecayed tissue by dark brown or black margins. Lesions around infected lenticels appear sunken, water-soaked, and circular in shape. Both types of lesion can turn hard, black, and dry in storage. Other bacteria often invade the lesions, causing a soft, foul-smelling decay. The amount of tuber soft rot is usually greater when certain decay fungi are present, particularly *Fusarium*.

Seasonal Development

Two variants or subspecies of the bacterium *Erwinia carotovora*, *E. c.* pv *atroseptica* and *E. c.* pv *carotovora*, cause blackleg and soft rot of tubers. The former is more active under cooler temperatures, below 70°F (18°C), and in some situations causes more of the inky black symptom associated with blackleg; the latter is more active under warmer temperatures, above 70°F, and sometimes causes more stem rot without the inky black symptom. The bacteria survive in infected tubers, on the roots of some other crops, including grains and sugar beets, on the roots of a number of weeds common in potato fields, including nightshades, lambsquarters, pigweeds, Russian thistle, kochia, purslane, and mallow, and in surface water used for irrigation.

Blackleg may be spread from infected to noninfected seed pieces during cutting. Infection is most likely if wound healing is delayed, usually because tubers are not warmed before cutting or because seed pieces are planted in cold, dry soil. Warm soil favors wound healing, but if the soil is wet, the bacteria may decay infected seed pieces before any sprouts emerge. If the soil is cool after planting, then infected seed pieces decay more slowly, and the bacteria move up into the water-conducting tissue of the stems sometime after sprouts emerge. The bacteria multiply and degrade the vascular tissues, causing foliar symptoms and wilt. Later, bacteria spread into stem pith tissue, causing the darkening and softening that are the typical blackleg symptoms.

Bacteria in surface irrigation water may be an important inoculum source in most areas. These bacteria can infect leaves, stems, and tubers or seed pieces through wounds or lenticels. Bacteria can also reside in lenticels without causing decay; these bacteria may cause blackleg if tubers are used for seed. Decaying tubers release large quantities of bacteria into the soil. The bacteria infect more tubers via soil water or may contaminate surface water. Spread of blackleg bacteria is favored by wet conditions. The bacteria can also be spread from cull piles to healthy plants by flies, but the importance of such spread is not known.

Wet soil conditions favor tuber infections; immature tubers are more susceptible to bacterial infection than are tubers with fully developed skin. Proper curing and low storage temperature prevent or restrict soft rot of tubers. Moisture leaking from soft-rotted tubers will spread infection in storage. Free moisture on tuber surfaces and poor storage ventilation restrict the oxygen supply to tubers and increase soft rot because the bacteria grow better under low oxygen.

Management Guidelines

Using seed stocks with low levels of blackleg is a fundamental part of managing seed piece decay and blackleg. Obtain seed tubers from a source you know provides good quality; seed originating from stem cutting or micropropagation programs is usually the best. However, poor handling or environmental stress after planting can result in high amounts of blackleg or seed piece decay, even when the best quality seed is used.

To reduce the incidence of seed piece infections, hold seed tubers at 50° to 55°F (10° to 13°C) with relative humidity at 95% for 10 to 14 days before cutting, if excessive sprouting is not a problem. Follow good sanitation practices during seed cutting and handling to reduce spread of bacteria. Wash equipment clean and apply a registered disinfectant (see Table 7) at least twice a day and always between seed lots.

Plant in well-drained, moist soil with a temperature of 50°F (10°C)or higher to encourage rapid wound healing. Colder and drier conditions slow wound healing, increasing the likelihood of infection. However, avoid excessive soil moisture, which also favors seed piece decay before emergence. When planting cannot be done in warm soils, treatment of seed tubers with approved fungicides may reduce seed piece decay; fungicides control *Fusarium*, which interacts with *Erwinia* to cause more serious seed piece decay.

Eliminate cull piles and control weeds and potato volunteers in rotation crops and adjacent fields. When growing seed potatoes, avoid using surface water for irrigation. In early field generations where blackleg incidence is very low, remove plants with blackleg or stem rot symptoms as soon as they appear. If you rogue blackleg plants, be sure to remove all tubers; place them in plastic bags as soon as you pull them from the soil to avoid spreading bacteria. Roguing is not practical during later generations when the number of blackleg plants is large.

Because immature tubers are infected more easily, allow tubers to mature before harvest. Avoid excessive soil moisture at harvest. Properly suberize tubers after harvest by holding them at 50° to 55°F (10° to 13°C) and 95% relative humidity for 10 to 14 days with good

Stems of plants with blackleg usually turn black just above and below the soil line. Plants may wilt and leaves may curl up and turn brown at their margins.

Bacterial soft rot lesions caused by *Erwinia* are separated from undecayed tissue by dark brown or black margins. The lesions may later turn hard, black, and dry.

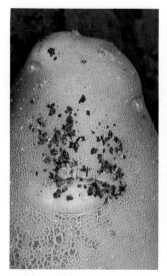

Rhizoctonia forms dark brown or black sclerotia, called black scurf, on tuber surfaces.

Rhizoctonia forms reddish brown lesions on underground parts of stems.

ventilation. Store tubers as cold as possible, considering their intended use, with good ventilation.

Rhizoctonia Stem and Stolon Canker
Rhizoctonia solani

Though widespread, Rhizoctonia stem and stolon canker causes only small yield reductions in the West unless cool, wet weather occurs during sprout growth. Otherwise, the principal damage caused by *Rhizoctonia* is loss in tuber quality. Planting seed tubers free of *Rhizoctonia* sclerotia and using cultural practices that favor rapid sprout emergence are the best control measures.

Symptoms and Damage

Rhizoctonia causes the most severe losses under cold, wet soil conditions during early sprout growth when it can kill sprouts before they emerge. This type of damage, uncommon in the West, results in reduced stands and large yield losses.

In the western states, damage is normally first seen after plants are established. Reddish brown lesions on stolons and underground parts of the stems are characteristic. Some infected stolons may be killed as they form. This stolon pruning results in fewer but larger tubers on each plant.

Sometimes *Rhizoctonia* lesions penetrate stem tissue far enough to interfere with nutrient flow in the conducting tissues. When this happens, the foliage may be wilted, stunted, rosetted, or curled upward. Leaves of white-skinned or russet varieties may turn yellow; leaves of red-skinned varieties may turn reddish or purple. Foliar symptoms may appear similar to those caused by leafroll virus. Aerial tubers may form at the base of stems, but aerial tubers may also be caused by blackleg, aster yellows, or waterlogging.

Sometimes a white fungal growth develops on stems just above the soil line. Under cool, wet conditions after tubers are formed, the fungus develops black or dark brown sclerotia called *black scurf* on tuber surfaces.

Seasonal Development

Many strains of *Rhizoctonia solani* exist and infect a wide variety of plants including grains, legumes, and vegetable crops. The strain that attacks potatoes does not cause disease on other plants. It survives in plant debris in the soil or as black scurf on the surface of tubers. For unknown reasons, the potato strain of *Rhizoctonia* does not survive well in the soil of many areas.

In the presence of growing stems or stolons, the fungus is stimulated to form infection structures. If the soil is wet and cool and the shoot tips are growing slowly, the fungus will completely invade and kill them. Maximum disease development occurs at soil temperatures below 54°F (12°C); infections and damage by *Rhizoctonia* decrease at higher temperatures. The fungus will form black scurf on tuber surfaces anytime the soil is wet and cool; most sclerotia form when tubers are left in the ground after vine death.

Management Guidelines

Good cultural practices that favor rapid emergence are the best control for Rhizoctonia. Avoid planting in wet soil. Warm the seed tubers before cutting, plant seed pieces shallowly when soil temperature is above 45°F (7°C), and form up hills after sprouts have emerged. Do not irrigate cold soil until sprouts have emerged.

Use crop rotations to keep *Rhizoctonia* from increasing to more damaging levels. Allow crop residues to decompose before planting; damage may be more severe following sugar beets, alfalfa, or clover if the crop residue is not decomposed before planting potatoes.

Control of seedborne inoculum is important in potato fields where the potato strain of *Rhizoctonia* does not survive in the soil. Avoid using seed tubers that have *Rhizoctonia* sclerotia on their surface; try to cull these during seed-cutting operations.

Bacterial Ring Rot
Corynebacterium sepedonicum

Ring rot is probably the most feared disease of potatoes; there is a zero tolerance for it in seed certification programs, meaning that one plant or tuber diagnosed as having ring rot symptoms causes an entire seed lot to be rejected for certification. Because of strict certification requirements, ring rot appears only occasionally in commercial fields. When it does occur it usually causes large

Ring rot causes interveinal tissue of leaflets to turn pale then brown. Affected stems wilt.

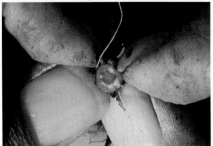

M.K.C. SUN

Bacterial slime may be squeezed from stems infected with ring rot bacteria.

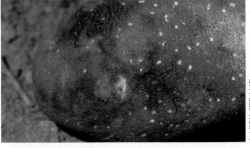

LARRY STRAND

The surface of a tuber decaying with ring rot turns pink or brown and cracks develop.

KEN KNUTSON

In Russet Burbank and a few other cultivars, the terminal leaflets of young stems may appear bunchy or rosetted if infected by bacterial ring rot.

The vascular ring of a tuber infected with ring rot usually turns brown and corky as it decays. Soft, decayed tissue can often be squeezed out of cross-sectioned tubers. Secondary invasion by other bacteria may cause complete disintegration of tuber tissue.

yield and storage decay losses. Ring rot is controlled by using certified seed and strict sanitation practices.

Symptoms and Damage

In a few varieties, including Russet Burbank and Nooksack, bunchy or rosetted terminal leaves may appear on young plants. These symptoms are sometimes called *early dwarf* or *green dwarf*. Plants with these symptoms usually develop more typical foliar symptoms later.

Foliar symptoms generally appear at midseason or later, usually on middle and lower leaves first. Pale green areas appear on leaf margins or as interveinal mottling and later turn brown, giving leaves a burned appearance. Symptoms move upward until the entire stem wilts and dies. Usually only one or two stems in a hill are affected. Stems sometimes wilt without any discoloration of leaves. In cooler areas, symptoms are less pronounced or may not be expressed; plants may wilt without dying. Symptoms are also masked when soil nitrogen levels are high.

Vascular tissue in stems may turn brown. If infected stems are cut near the seed piece, a milky exudate can usually be squeezed from the cut end.

The characteristic symptom of ring rot is a decay of the vascular ring in tubers. This decay usually appears after foliar symptoms, but does not always develop in the tubers of infected plants. Tuber decay usually starts as a softening and creamy yellow discoloration of the vascular ring that can be squeezed out after tubers are cross-sectioned. The vascular ring usually turns brown as it decays; browning of the vascular ring sometimes occurs without the soft decay.

Tuber surfaces may become slightly pink or brown, and cracks that penetrate to the rotted vascular ring may appear on the surface. Secondary invasion by other bacteria may cause a complete breakdown of the middle of the tuber with little or no discoloration of the vascular ring. Secondary breakdown can occur in the field, in transit, or in storage.

Under cool conditions plants may develop few symptoms while producing infected tubers. Infected tubers can be symptomless at harvest and take up to 2 or 3 months to develop ring rot in storage; sometimes symptoms do not develop in infected tubers.

Seasonal Development

The bacterium that causes ring rot, *Corynebacterium sepedonicum*, overwinters in infected tubers. It does not live freely in the soil, but it can survive for several years as a dried slime on harvesting and grading machinery, sacks, and surfaces of storage and transport facilities. *Corynebacterium* is highly contagious; the bacterium infects tubers through wounds that reach into the vascular ring. The principal means of spread is seed cutting; a knife that has cut one infected tuber can spread the bacteria to at least 20 more cut surfaces. Picker planters will also spread ring rot bacteria during planting.

The ring rot bacterium is a vascular parasite. As stems grow from an infected seed piece, bacteria move up in the water-conducting tissues (xylem), multiply, and produce toxins that cause the foliar symptoms. Rosetting and other early dwarf symptoms occur when bacteria proliferate in very young stems of certain varieties.

About one-half to three-fourths of the daughter tubers of an infected plant will be infected with ring rot bacteria. Not all daughter tubers develop ring rot symptoms. Ring rot develops in tubers most rapidly at 64° to 72°F (18° to 22°C), and only slightly at 37°F (3°C). Ring rot normally does not spread from tuber to tuber in storage, but pockets of soft rot affecting several tubers may develop around tubers that have disintegrated from ring rot.

Management Guidelines

Use only certified seed tubers. Do not use tubers for seed if they came from an infected lot or a lot that was suspected of having any ring rot. Use only seed tubers free from bacterial ring rot for seed production.

Corynebacterium can overwinter in tubers left in the ground, so avoid planting potatoes on ground that produced infected potatoes the previous year. Leaving a field fallow or rotating it for at least 1 year should eliminate any ring rot bacteria present, as long as volunteers are controlled. Destroy potato volunteers that appear near seed fields.

Clean and disinfect all equipment, bins, containers, sacks—everything that touches tubers. Sanitation is most important during cutting and planting. Keep all seed lots separated from each other and from other potatoes. Clean and disinfect equipment between the handling of each lot or cultivar. All surfaces must be cleaned of debris with detergent and water or steam, and treated with a disinfectant registered for ring rot control. Even equipment that has been exposed to the weather and has not touched potatoes for several years must be disinfected. Be sure to use correct disinfectant concentrations and treatment times (see Table 7).

Disinfect storage facilities. Remove all sacks and debris; if storages have dirt floors, remove 2 or 3 inches of soil before disinfecting. Burn contaminated sacks.

If a commercial field shows ring rot, leave tubers in the ground as long as possible to allow most of the infected tubers to disintegrate completely. Harvest the remaining tubers and market them immediately. If you must store them, store them at temperatures as low as possible depending on their intended use, and sell them as soon as possible.

Scab
Streptomyces spp.

The causal organism of scab, a widespread soilborne disease, can infect fleshy roots and underground stems of some other crops and weeds, but causes economic damage only on potatoes and red beets. The surface blemishes it causes do not reduce yields but can reduce the value of a potato crop considerably. Severity of scab can be reduced by maintaining adequate soil moisture during early tuber development, using certain soil treatments, and avoiding heavily infested soils.

Symptoms and Damage

The surfaces of tubers infected with scab have brown, roughened, irregularly shaped areas that may be raised and warty, level with the surface, or sunken. Scab lesions may affect just a small part of the tuber surface or almost completely cover it. Three types of symptoms occur: *russet scab* produces lesions that are superficial, corky areas; *raised scab* causes erupting or cushionlike lesions; *deep-pitted scab* produces dark brown to almost black lesions that are sunken up to 1/4 inch (6 mm) deep. The type and severity of symptoms are influenced by the strain or species of scab organism. Tubers with raised or russet scab can be used for processing; tubers with deep-pitted scab cannot. Scab tends to be more severe in newly cropped fields and fields high in undecomposed organic matter.

Seasonal Development

Scab is caused by species of the soil microorganism *Streptomyces*. *Streptomyces scabies* causes russet and raised scab; different species cause deep-pitted scab. *Streptomyces* survives in soil in the absence of host plants, and can attack the fleshy roots of weeds and root crops including red beets, sugar beets, and carrots. The microbe invades potato tubers through lenticels during the first 5 weeks of tuber development. If tubers dry out during this period, bacteria antagonistic to *Streptomyces* that are normally present in the lenticels disappear, allowing the scab organism to infect more easily. A soil pH of 7 is optimum for scab development; a pH lower than 5.5 usually inhibits it. There is a type of scab called *acid scab* that develops at a pH below 5; it occurs on potatoes in the eastern United States, but is not known to occur in the western states.

As tubers grow, lesions gradually develop into characteristic scabs. The type of lesion formed depends on the tolerance of the potato cultivar, the species of *Streptomyces* involved, and the soil environment. Lesions do not continue to develop in storage.

Management Guidelines

Potato cultivars differ in their resistance to scab. Russet cultivars tend to be more resistant than smooth-skinned cultivars. Russet Burbank is generally considered tolerant. The relative tolerance of cultivars to scab varies with the area in which they are grown, because of differences in the strains of the disease organisms present and, possibly, environmental effects. If you are planting in fields where scab is a problem, ask local extension agents, farm advisors, or other experts for cultivar recommendations.

Proper soil moisture during tuber development reduces the severity of scab, and usually controls the disease adequately. Maintain available soil moisture at 80% of field capacity or more during tuber initiation and early tuber growth, until tubers are golf-ball sized. Avoid planting potatoes in fields with severe scab problems. Preplant treatment of such fields with materials designed to lower soil pH or control *Streptomyces*, combined with proper water management, may reduce the amount of scab. Water management and soil treatments may not be effective controls for deep-pitted scab in some areas; if this is the case in your area, you may not be able to grow potatoes profitably in fields where this type of scab occurs.

Fusarium Dry Rot
Fusarium spp.

Losses from dry rot occur wherever potatoes are stored, but can usually be kept to a minimum with proper handling and storage techniques.

The first signs of dry rot are small brown areas that appear around wounds about 1 month after tubers are put into storage. The skin over the dry-rotted areas sinks and becomes wrinkled; internally, lesions are light to dark brown, dry and spongy in texture, and tend to form hollow cavities. If dry rot lesions are invaded by bacteria, they become slimy and foul smelling. Pockets of rotting tubers frequently develop around dry-rotted tubers invaded by bacterial soft rot. Dry rot takes several months to develop fully.

Fusarium roseum and *Fusarium solani* are the primary soil fungi involved in dry rot. They are present in all soils and survive for many years as resistant spores; infections are caused by spores carried in soil on tuber surfaces. *Fusarium* cannot penetrate intact tuber skin, lenticels, or suberized seed pieces; wounds are required for infection. Scab or blight lesions, bruises, and cuts are common entry sites. In the presence of a tuber wound, fungus spores germinate and enter the tuber, where the fungus slowly decays the tissue, causing typical dry rot symptoms unless bacterial soft rot develops. Damp conditions in storage favor soft rot development.

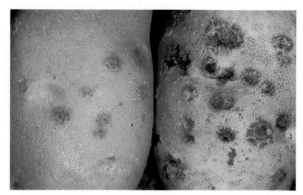

Scab lesions may appear as eruptions, pits, or superficial layers of corky tissue.

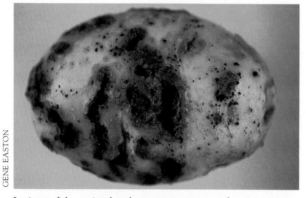

Lesions of deep-pitted scab are more severe and penetrate more deeply than other types of scab lesions.

GENE EASTON

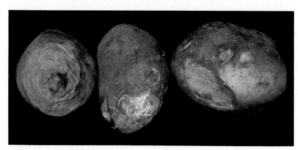

The surface of Fusarium dry rot lesions becomes sunken and wrinkled. White or pink mycelium may develop on the surface.

Proper handling and curing is usually sufficient to give economic control of dry rot in storage. Allow tubers to mature before harvest and prevent bruising or wounding of tubers during harvest. Wound healing reduces infection by *Fusarium*; for the first 2 to 3 weeks of storage, hold tubers at 50°F (10°C) with good ventilation and a relative humidity of at least 95%. Fungicide treatments before storage may be recommended when tubers must be harvested under adverse conditions or when conditions are not favorable for adequate wound healing.

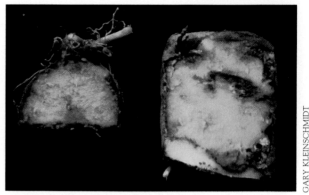

GARY KLEINSCHMIDT

Seed piece decay caused by Fusarium appears as reddish brown lesions that may cover the entire seed piece. Decayed seed pieces may turn black and slimy if invaded by bacteria.

Seed Piece Decay
Fusarium spp. and *Erwinia carotovora*

Seed piece decay occurs wherever potatoes are grown. It is most severe when unfavorable conditions occur after planting. Seed piece decay is managed with proper seed handling and planting procedures and seed-treatment fungicides when necessary.

The first signs of seed piece decay are reddish brown to black spots that slowly form depressions on the surfaces of cut seed pieces. Later these spots may become black and slimy from bacterial infections. Entire seed pieces will rot if enough infections develop. Seed piece decay reduces stands, results in variable plant size, and may increase the incidence of blackleg.

Seed piece decay is usually caused by the combined action of the same *Fusarium* fungi that cause dry rot and the same *Erwinia* bacteria that cause blackleg, although sometimes decay can be caused by *Fusarium* alone. *Fusarium* spores are present virtually everywhere in the soil, in storage, and on handling equipment. They germinate and infect the freshly cut surfaces of seed pieces, causing lesions that later may be invaded by *Erwinia*. When seed pieces are infected by both *Erwinia* and *Fusarium*, losses from seed piece decay are much greater than when *Fusarium* is present by itself. Cool soil that is too wet or too dry favors the development of seed piece decay.

Seed piece decay can be controlled adequately when seed pieces can be planted under conditions that favor rapid suberization; *Fusarium* cannot infect cut surfaces after they are suberized. Warm the seed tubers to 50°F (10°C) before cutting, and keep cutting and handling equipment disinfected. Plant when the soil temperature is at least 45°F (7°C) and when soil moisture is 60% to 80% of field capacity. If possible, avoid irrigating before emergence. When planting conditions are likely to favor seed piece decay, treat cut seed pieces with a registered fungicide.

Pink Rot
Phytophthora erythroseptica

Pink rot is a tuber decay that occurs sporadically in many of the growing areas in the West. It is most important in parts of the Klamath Basin growing area, where it is a serious tuber decay problem. The causal organism, *Phytophthora erythroseptica*, causes disease on potatoes, tulips, and certain woody vines and ornamentals. The fungus survives for long periods in many soils and becomes active when the soil is nearly or completely saturated with water. The disease is much worse when saturated soil is accompanied by warm temperatures. The only control is to avoid poorly drained soils and saturated soil conditions, especially at or near harvest.

Phytophthora erythroseptica infects roots and stolons as well as tubers, and sometimes causes a wilting and early dying of plants that may be confused with Verticillium wilt. The most important phase of the disease is tuber decay. From infected eyes or lenticels, rot spreads through the tuber. When the infected tuber is cut, the rotted portion is delineated by a dark line at its margin. The decayed tissue is spongy but not discolored. With exposure to air, the surface of the decay turns a salmon pink color in about half an hour, turning to brown and then black after about an hour. Tuber infections can spread in storage when spore-containing liquid oozes from decaying tubers; the liquid can also establish conditions favoring bacterial soft rot. To prevent rot in storage, avoid harvesting wet tubers whenever possible, maintain good airflow and avoid the accumulation of moisture on tubers, and keep the temperature as low as possible. The fungus is inactive below 40°F (4.4°C).

Water Rot
Pythium spp.

The fungus *Pythium aphanidermatum* causes water rot in Arizona. The disease is similar to pink rot tuber decay; the causal organisms are similar and control measures are essentially the same. In certain areas of Arizona, *P. aphanidermatum* can cause a severe stem rot similar to blackleg when fields are overirrigated late in the season. Total losses from tuber decay may occur if infected fields are irrigated after vine death and before harvest. Limiting postharvest irrigations to avoid wet soil late in the season eliminates these problems.

In the Columbia Basin, a different species of *Pythium*, *P. ultimum*, causes tuber decay. The disease tends to occur on tubers from fields planted to potatoes for the first time, then disappears after one or two crops of potatoes. Symptoms occur after tubers are in storage, when they shrivel and dry up. The fungus infests the tuber surface in the field, but does not infect the tuber unless short periods of dryness occur during transit or storage. The disease does not spread in storage. Although *P. ultimum* is closely related to *P. aphanidermatum*, the tuber decay it causes is not increased by late-season irrigations. To control decay caused by *P. ultimum*, maintain adequate humidity throughout the storage period.

Black Dot
Colletotrichum atramentarium

The economic significance of black dot, a widespread potato disease in western growing areas, is unknown. The disease also occurs on other plants in the potato family, including tomatoes, eggplants, and peppers, and may play an important role in other potato diseases.

The black dot fungus, *Colletotrichum atramentarium*, can cause severe rot of all underground plant parts; in some cases belowground symptoms might be confused

Tuber tissue decaying with pink rot has a dark margin, but is not discolored when the tuber is first cut (*center*). The cut surface of a pink rot lesion turns salmon pink in about half an hour (*left*), then black after about an hour (*right*).

The black dot fungus forms a thin layer of white mycelium that produces tiny, black, dotlike structures called stroma on the stems of dying plants.

with Rhizoctonia stem and stolon canker. Yellowing and wilting of foliage may occur, starting at the tops of plants and moving downward. Wilt develops rapidly, in contrast to Verticillium wilt. Black dot is distinguished by small, black, dotlike fungal structures (stroma), that form on collapsing stems and on tubers. Another common symptom of black dot is the adherence of stolons to the stem ends of tubers. *Colletotrichum* often invades plants that are weakened by other diseases, and may accelerate early death of vines infected with *Verticillium*, *Erwinia*, and possibly *Phytophthora*. Black dot occurs most frequently on plants grown in coarse-textured soils, under conditions of low or excessively high nitrogen, high temperature, or poor soil drainage.

Colletotrichum survives as stroma that form on tubers, stolons, roots, and stems at the end of the season, when plants are mature and starting to collapse. The pathogen is thought to be spread on contaminated seed tubers, and increases in fields that are continuously cropped to potatoes. For control of black dot, maintain adequate soil fertility, avoid excess water, and plant grains for 2 or 3 years before replanting potatoes on infected land.

Silver Scurf
Helminthosporium solani

Silver scurf occurs in most potato-growing areas, but is of minor importance. The disease affects only the tubers of potato plants; no other hosts are known. Silver scurf increases weight loss during storage and reduces market quality.

Symptoms of silver scurf usually appear at the stem end as small, light brown or grayish, leathery spots. The spots have a shiny, silvery appearance, which is more pronounced when tubers are wet, and they may enlarge to cover much of the tuber surface. The margins of young lesions frequently appear dark brown and sooty from spore production. Silver scurf is most noticeable on red-skinned cultivars. Because the lesions disrupt the periderm, affected tubers are more susceptible to decay and shrinkage from water loss during storage.

When infected seed pieces are planted, silver scurf lesions develop on the seed pieces. Spores produced on these lesions are dispersed into the adjacent soil, where they are able to infect other tubers for at least 6 months. Clean seed pieces may be infected if planted in infested soil the following year. *Helminthosporium solani* infects tubers through lenticels or directly through the skin. The fungus remains confined to the periderm and the outer layers of cortex cells. Most spores are produced on the margins of young, developing lesions. For this reason, infected seed pieces with small lesions tend to produce plants with more infected daughter tubers. Silver scurf can spread in storage; the fungus will develop at temperatures above 37°F (3°C) and humidity above 90% to 93%.

For control of silver scurf, plant certified seed tubers, harvest tubers as soon as they have matured after vine killing, use humidified airflow to remove free moisture from wet tubers as quickly as possible after harvest, and store tubers at temperatures as low as possible. Avoid planting potatoes in a field that had silver scurf the previous season.

Stem Rot
Sclerotium rolfsii

The fungus *Sclerotium rolfsii* attacks many crops, including alfalfa, tomatoes, carrots, beans, and potatoes in warm climates. In the West, the disease causes significant potato losses in some areas of Kern County, California, rotting tubers in the field or in transit. Stem rot can be controlled with cultural practices or fumigation of heavily infested soil.

Sclerotium rolfsii survives in the soil as poppyseed-sized, dark brown to black sclerotia that form on infected stems and tubers. Sclerotia germinate when the weather is warm, and the fungus invades dead or dying stem tissue that is at or below the soil surface. *Sclerotium* is most active when the temperature is about 80° to 90°F (27° to 32°C), growing over tubers as a thin, white, cobwebby layer, and penetrating into tubers through lenticels or from dead stolons. Stem rot appears near the end of the season, sometimes after vines are completely dead, and may destroy tubers before harvest. Infected tubers may decay in transit or in storage.

Planting winter cereal crops followed by summer fallow may reduce the amount of fungus in infested fields. Avoid planting susceptible crops in infested fields. If you plant potatoes in fields known to be infested with *Sclerotium*, plant early to avoid late-season high temperatures that favor stem rot. Soil fumigation may control *S. rolfsii* for at least 2 years. Application of ammonium bicarbonate in the last one or two irrigations before harvest has kept stem rot in check, however no specific recommendations are presently available. Contact your extension agent, farm advisor, or other experts for the latest recommendations.

Powdery Scab
Spongospora subterranea

A disease of minor importance, powdery scab occurs in areas of Washington, Oregon, and California. The disease is favored by cool, damp conditions, is found mainly on coarse-textured soils, and is more severe on white-skinned potato varieties. The fungus that causes powdery scab also infects nightshades.

Spongospora subterranea is a slime mold fungus that infects lenticels, wounds, or eyes of tubers, and stolons and roots. Scablike, warty lesions form on tuber surfaces and fill with dark brown, powdery masses of spores. Lesions remain superficial unless the soil is very wet, when they may form larger, deeper cankers.

The fungus survives for several years as resting spores in the soil or on tuber surfaces. In the presence of susceptible roots the spores germinate and infect the surface cells of roots or stolons. Symptoms on roots may be confused with root-knot nematode symptoms. These infections produce a second kind of spore that swims through free water in the soil and infects root, stolon, or tuber surfaces.

For control of powdery scab, plant certified seed tubers, do not overirrigate, and avoid planting potatoes in contaminated fields. The fungus may be reduced by growing crops other than potatoes for several years and by controlling nightshades. Russet cultivars are more resistant to powdery scab.

FOLIAR DISEASES

Important foliar diseases of potatoes include early blight, late blight, powdery mildew, and white mold. Foliar diseases are caused by fungi that are dispersed by wind or water and infect potato foliage, forming lesions on leaves or stems. Early and late blights can also infect tubers. Because the pathogens involved can be dispersed by wind, these diseases can spread rapidly. Careful monitoring is required in areas where these diseases may develop, because timing of control measures is usually critical. In certain areas foliar fungicides are sometimes needed to control early blight, and timing is based on degree-day accumulations or observations of disease development. To stop the spread of late blight, fungicides must be applied as soon as symptoms begin to develop or when weather conditions favorable for disease occur. White mold is controlled by increasing intervals between irrigations when the disease appears, to allow the canopy to dry out.

Early Blight
Alternaria solani

Early blight can affect both foliage and tubers wherever potatoes are grown. If not controlled, it can reduce both tuber yield and quality. Early blight is most serious in Rocky Mountain growing areas and parts of the Snake River Valley. The disease is most severe under alternately wet and dry conditions, consequently the increased incidence of early blight is attributed to the increased use of sprinkler irrigation. Proper management of soil moisture and fertility decreases the severity of early blight. Foliar blight is controlled with properly timed

Silver scurf appears as light brown or grayish leathery or russeted areas on the surface of red- or white-skinned tubers.

Sclerotium rolfsii forms thin, white, cobwebby mycelium on stems and tubers at the end of the season.

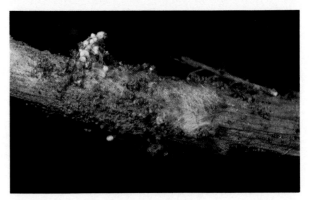

Sclerotia formed by *Sclerotium rolfsii* are poppyseed sized and white at first, turning dark brown to black.

fungicide applications, and tuber infections are reduced by delaying harvests. Early blight also occurs on tomatoes and other plants of the potato family.

Symptoms and Damage

Early blight is primarily a disease of stressed or aging (senescing) plants. Early blight infections are characterized by brown lesions that appear first on the oldest (lower) leaves. The lesions are usually circular, about 1/8 to 3/8 inch (3 to 10 mm) in diameter and consist of concentric rings of dead tissue that give a targetlike or bull's-eye appearance. Lesions become angular in shape when expansion is limited by leaf veins. A yellow zone that fades into the green of healthy tissue often surrounds the lesion. Affected leaves usually do not fall off.

As the disease begins to spread, lesions appear higher up on the younger leaves of the plant. Dark brown to black stem lesions may also occur at this time. Early blight symptoms are more severe on vines dying from natural aging or stressed by other diseases such as Verticillium wilt. If early blight foliage infections become severe, harvest yields can be reduced significantly. Tuber infections that occur during harvest can result in significant storage losses and reduced tuber quality.

Tubers are infected at harvest by *Alternaria* spores deposited on the soil during the foliar phase of the disease. Mature tubers usually must be wounded during harvest for infection to occur, and free moisture must be present on the tuber surface. Therefore, tubers grown in coarse, sandy soil and tubers harvested when wet are most likely to be affected. Tuber lesions appear during storage, usually several months following harvest. Lesions are small, brown to black, with surface margins that are purple to black in color. Where they penetrate into the tuber flesh, they are brown and slightly leathery, and they usually remain small and superficial under good storage conditions. Tuber lesions reduce the market value of the crop. In parts of Idaho, Colorado, and other Rocky Mountain growing areas, tuber lesions are the most serious phase of the disease, particularly with the smooth, white-skinned cultivars commonly grown for chipping.

Seasonal Development

The early blight fungus, *Alternaria solani*, overwinters on potato refuse in the field, in soil, on tubers, and on other plants of the potato family. Spores produced by the fungus come in contact with leaves touching the soil or are carried to leaf surfaces by wind. In the presence of free moisture, spores germinate and the fungus penetrates directly through leaf surfaces. The fungus grows within the plant for a time without causing symptoms; the first lesions usually appear during early tuber growth, about

1 week after flowering. Symptoms are more pronounced on leaves that are weakened by age, stress, poor nutrition, or diseases such as Verticillium wilt. In some areas, early blight does not occur until late in the season.

Secondary spread occurs when spores produced on leaf lesions are carried to other sites by air movement. The most rapid spread of early blight occurs in alternating wet and dry weather. Moist foliage conditions created by heavy dews, frequent rains, sprinkler irrigation, or high humidity are necessary for spore germination and infection. However, dry conditions favor wind transport of spores from leaf lesions to other foliage. Appearance of early blight lesions is an indication that secondary spread is occurring. Spore trapping may be used to detect secondary spread of *Alternaria*. Spore traps generally detect spread before lesion development is readily apparent in the field.

Tubers become infected as they are lifted through soil infested with *Alternaria* spores, which tend to be concentrated at the soil surface. Conditions that increase the severity of foliar blight will increase the quantity of spores on the soil surface at harvest. Because tuber infections occur mostly through wounds, immature tubers and tubers of white- and red-skinned cultivars are infected more readily. Harvesting tubers from coarse, sandy soils can also increase disease severity by increasing injury. Letting tubers mature after vine death allows proper skin development and decreases the amount of tuber injury and tuber infection. Harvesting when vines are green increases tuber infections because tubers are immature. Infections can also occur through lenticels, which are open when wet; therefore, tuber infection is increased by harvesting wet tubers.

Management Guidelines

In many growing areas, appropriate cultural practices are usually sufficient to avoid economic loss from early blight. Practices that optimize growing conditions through proper fertilizer, water, and pest management reduce stress and, therefore, decrease the severity of early blight. Generally, late-maturing cultivars are less susceptible to the foliar phase of the disease. However, if tubers are not allowed to mature properly before harvest, tuber infection can still be severe.

Nitrogen and phosphorus deficiencies increase susceptibility to early blight. Use petiole analysis as a guide to maintain levels of nitrogen and phosphorus considered adequate for your growing area. Make sure potassium levels are adequate; otherwise tuber skin will not develop properly and tuber infection may increase.

Reduce early blight infections of tubers by letting tubers mature in the ground for at least 2 weeks after vine death. Avoid harvesting wet tubers; if conditions force you to harvest wet tubers, dry them as quickly as possible with forced ventilation as soon as they are placed in storage. Follow procedures that minimize mechanical injury during harvesting and handling. In all situations, conditions that favor rapid wound healing during the first 2 weeks of storage will reduce the severity of tuber infection.

Application of foliar fungicides is economically justified in areas where the disease becomes severe before vine death begins from other causes or where tuber blight is a problem. Early blight controls are presently recommended in some of the Klamath Basin, Snake River Valley, and Rocky Mountain growing areas. To be effective, applications must be properly timed and the fungicides must thoroughly cover the vines. Foliar sprays should be applied when secondary spread of early blight begins. Degree-day models, spore trapping, and field monitoring may be used to determine when secondary spread begins and spray programs should be started. Guidelines used in Colorado and Idaho for timing early blight sprays are given below. Sprays for early blight control have a greater effect on tuber yield when other diseases that can mask the beneficial effects of early blight control, such as Verticillium wilt, are controlled. Fungicide applications are not always economically justified, because tuber blight may not be a problem and foliar blight may be mild or occur too late in the season to cause significant loss. Consult your local extension agent, farm advisor, or other experts for specific recommendations concerning your area.

Using Degree-days in Colorado. A degree-day model has been developed for scheduling early blight sprays in Colorado, where early blight is a problem every year if not controlled. The model estimates the time of first lesion appearance and, therefore, secondary spread of early blight by using heat units calculated as *degree-days*. The time of first lesion appearance depends on the amount of heat the plant and pathogen experience, so measuring the amount of heat accumulating over time is biologically more meaningful and more accurate than using calendar days to predict outbreaks. A degree-day is the amount of heat that accumulates during a 24-hour period when the average temperature is one degree above the developmental threshold for the pest or plant in question. The Colorado early blight model accumulates degree-days as degrees Fahrenheit and uses a developmental threshold of 45°F (7°C).

To use degree-days for scheduling treatments, start recording daily minimum and maximum temperatures in your field after the earliest planting in your area. Use Table 12 to determine the total degree-days for each day. After planting, begin to tally degree-days for the season using a form such as the one in Figure 34. To properly

MINIMUM TEMPERATURES

MAX TEMPS	70	68	66	64	62	60	58	56	54	52	50	48	46	44	42	40	38	36	34	32	30	28	26	24	22	20
120	50	49	48	47	46	45	44	43	42	41	40	39	38	37	36	36	35	34	34	33	33	32	32	31	31	30
118	49	48	47	46	45	44	43	42	41	40	39	38	37	36	35	35	34	33	33	32	32	31	31	30	30	30
116	48	47	46	45	44	43	42	41	40	39	38	37	36	35	34	34	33	32	32	31	31	30	30	29	29	29
114	47	46	45	44	43	42	41	40	39	38	37	36	35	34	33	33	32	31	31	30	30	29	29	28	28	28
112	46	45	44	43	42	41	40	39	38	37	36	35	34	33	32	32	31	30	30	29	29	28	28	27	27	27
110	45	44	43	42	41	40	39	38	37	36	35	34	33	32	31	31	30	29	29	28	28	27	27	27	26	26
108	44	43	42	41	40	39	38	37	36	35	34	33	32	31	30	30	29	28	28	27	27	26	26	26	25	25
106	43	42	41	40	39	38	37	36	35	34	33	32	31	30	29	29	28	27	27	26	26	25	25	25	24	24
104	42	41	40	39	38	37	36	35	34	33	32	31	30	29	28	28	27	26	26	25	25	24	24	24	23	23
102	41	40	39	38	37	36	35	34	33	32	31	30	29	28	27	27	26	25	25	24	24	24	23	23	22	22
100	40	39	38	37	36	35	34	33	32	31	30	29	28	27	26	26	25	24	24	23	23	23	22	22	21	21
98	39	38	37	36	35	34	33	32	31	30	29	28	27	26	25	25	24	23	23	22	22	21	21	21	20	20
96	38	37	36	35	34	33	32	31	30	29	28	27	26	25	24	24	23	23	22	22	21	21	20	20	20	19
94	37	36	35	34	33	32	31	30	29	28	27	26	25	24	23	23	22	22	21	21	20	20	19	19	19	18
92	36	35	34	33	32	31	30	29	28	27	26	25	24	23	22	22	21	21	20	20	19	19	18	18	18	17
90	35	34	33	32	31	30	29	28	27	26	25	24	23	22	21	21	20	20	19	19	18	18	18	17	17	17
88	34	33	32	31	30	29	28	27	26	25	24	23	22	21	20	20	19	18	18	17	17	17	16	16	16	16
86	33	32	31	30	29	28	27	26	25	24	23	22	21	20	19	19	18	18	17	17	16	16	16	15	15	15
84	32	31	30	29	28	27	26	25	24	23	22	21	20	19	18	18	17	17	16	16	15	15	15	14	14	14
82	31	30	29	28	27	26	25	24	23	22	21	20	19	18	17	17	16	16	15	15	14	14	14	13	13	13
80	30	29	28	27	26	25	24	23	22	21	20	19	18	17	16	16	15	15	14	14	14	13	13	13	12	12
78	29	28	27	26	25	24	23	22	21	20	19	18	17	16	15	15	14	14	13	13	12	12	12	12	11	11
76	28	27	26	25	24	23	22	21	20	19	18	17	16	15	14	14	13	13	12	12	12	11	11	11	11	10
74	27	26	25	24	23	22	21	20	19	18	17	16	15	14	13	13	12	12	12	11	11	11	10	10	10	10
72	26	25	24	23	22	21	20	19	18	17	16	15	14	13	12	12	12	11	10	10	10	9	9	9	9	9
70	25	24	23	22	21	20	19	18	17	16	15	14	13	12	11	11	10	10	10	9	9	9	8	8	8	
68		23	22	21	20	19	18	17	16	15	14	13	12	11	10	10	9	9	9	8	8	8	8	7	7	7
66			21	20	19	18	17	16	15	14	13	12	11	10	9	9	9	8	8	7	7	7	7	7	6	
64				19	18	17	16	15	14	13	12	11	10	9	8	8	8	7	7	7	6	6	6	6	6	6
62					17	16	15	14	13	12	11	10	9	8	7	7	6	6	6	5	5	5	5	5	5	
60						15	14	13	12	11	10	9	8	7	6	6	5	5	5	5	4	4	4	4		
58							13	12	11	10	9	8	7	6	5	5	4	4	4	3	3	3	3	3	3	
56								11	10	9	8	7	6	5	5	4	4	4	3	3	3	3	3	3		
54									9	8	7	6	5	4	4	3	3	3	3	2	2	2	2	2	2	
52										7	6	5	4	3	3	2	2	2	2	2	1	1	1	1		
50											5	4	3	2	2	2	1	1	1	1	1	1	1	1		
48												3	2	1	1	1	1	1	1	1	1	0	0	0	0	

Table 12. Degree-Days Above 45°F, Based on Daily Maximum and Minimum Temperatures. To find the degree-days for a day, follow the column and row of the day's minimum and maximum temperatures to where they intersect. For odd-numbered temperatures, interpolate between the table values.

DATE	DAILY LOW	DAILY HIGH	DEGREE-DAYS FOR DATE	TOTAL DEGREE-DAYS FOR SEASON TO DATE
5/17	35°F	53°F	0	0
5/18	25	59	0	0
5/19	31	68	4.5	4.5
5/20	36	70	8.0	12.5
5/21	35	71	8.0	20.5
5/22	37	73	10.0	30.5
5/23	37	68	7.5	38.0
5/24	35	66	5.5	43.5
5/25	22	61	0	43.5
5/26	25	67	1.0	44.5

Figure 34. Sample form for recording degree-days in Colorado. Record the daily high and low temperatures and the degree-days for the date. Use Table 12 or calculate the degree-days as average daily temperature (in degrees Fahrenheit) minus 45, recording negative numbers as zeroes. The total degree-days for the season to date will tell you when to schedule the first spray application for early blight.

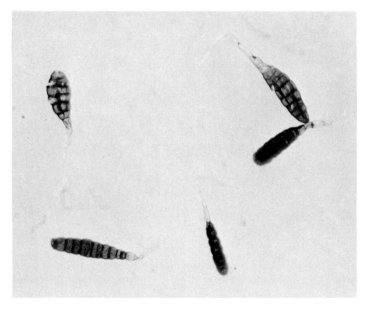

Figure 35. The large, dark, segmented spores of *Alternaria* are easy to identify with a compound microscope.

calibrate the model for a given production area, simultaneously observe lower leaves for the time of first lesion appearance. Begin looking for lesions before the first flowers form. Spore-trapping data can also be used to determine when secondary spread is occurring.

In the San Luis Valley, apply the first early blight spray when the degree-day total reaches 650°D; in northeastern Colorado, make the first blight application when the degree-days total reaches 1,125 to 1,150°D. The model is used only to time the initial application properly. Make subsequent applications according to label directions. Degree-day models such as this may be used for other areas where early blight requires control, but they must be calibrated for several years in each area to ensure their accuracy.

Spore Trapping. Trapping and identification of *Alternaria* spores can be used to determine when early blight is beginning to spread within an area. Spores are usually detected with spore traps before early blight is readily detected in the field. To use spore trapping, mount greased microscope slides in a weather-vane spore trap above the level of the potato canopy in a potato field. A weather station is a convenient location. Begin trapping about the time tuber initiation begins. Replace the slides with fresh ones at least twice a week and examine the exposed slides with a microscope to see if *Alternaria* spores are present; the spores are easily recognized (Figure 35). The presence of spores indicates that secondary spread of the pathogen is occurring. In areas where early blight is a perennial problem, you may choose to begin fungicide applications when spores appear; in other areas, begin monitoring fields for early blight symptoms at this time.

Spore trapping may be used to develop a degree-day model. If the appearance of *Alternaria* spores occurs consistently from season to season in a given area once a certain number of degree-days have accumulated, then that degree-day total can be used to make early blight management decisions.

Field Monitoring, Idaho Guidelines. In the upper Snake River Valley growing areas of Idaho (see Figure 1), apply fungicides at the first sign of secondary spread, as indicated by the appearance of target spot lesions in the upper two-thirds of the plant. Careful monitoring is required. The number of sprays recommended depends on the time between the beginning of secondary spread and vine killing. These recommendations may be adaptable to other areas where early blight may cause significant losses, such as the Klamath Basin and other California and Rocky Mountain growing areas.

- Do not spray until blight lesions appear on leaflets above the lower one-third of the plant.
- If the first spray is near July 9, follow with three more sprays 10 to 12 days apart.

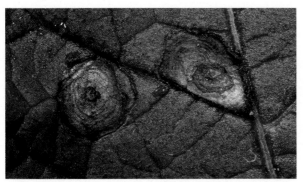

Early blight lesions are dark brown, with concentric rings that have a target appearance. They become angular in shape when limited by leaf veins.

Dark brown early blight lesions develop on stems in later stages of the disease.

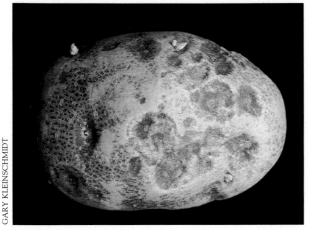

GARY KLEINSCHMIDT

Early blight lesions on tubers appear sunken and are brown or black. They usually remain superficial.

Late blight lesions on leaves are brown to purplish black, with pale green margins. White mycelium and spores may develop.

- If the first spray is near July 21, follow with two more sprays 10 to 12 days apart.
- If the first spray is near August 1, follow with one more spray 10 to 12 days later.
- If the first spray is near August 10, check closely to see if another spray is needed.
- If early blight builds up after August 15, do not spray in eastern Idaho; one spray might be applied in south-central Idaho, depending on harvest date.

Late Blight
Phytophthora infestans

Late blight is probably the most important disease of potatoes worldwide. Climatic conditions limit the importance of late blight in the western United States. It occurs regularly in the coastal valley growing areas; elsewhere its occurrence is sporadic, depending on the presence of the pathogen and cool, damp weather conditions. Late

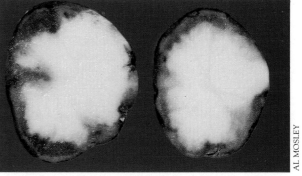

AL MOSLEY

Late blight tuber lesions consist of brown, granular, rotted tissue that extends a short distance into the tuber.

blight causes losses in potatoes as both foliar and tuber disease, and also affects other plants of the potato family, such as tomatoes, eggplants, peppers, nightshades, and groundcherries. Late blight is controlled by eliminating cull piles and volunteer potatoes, using proper harvesting and storage practices, and applying fungicides when necessary.

Symptoms and Damage

Lesions appear on leaves as small, water-soaked spots, usually at the tips or margins. Under the cool, moist conditions that favor late blight development, lesions grow to large, brown to purplish black lesions with pale green halos of water-soaked tissue at their margins. In humid or wet weather, a white growth of mycelium and spores appears at the edge of lesions, mostly on the underside of leaves. Lesions may kill entire leaflets and spread over the whole plant. Greasy, dark lesions may appear on stems, causing severe wilting or sometimes girdling and killing the plant. As the disease is spread through a field, characteristic teardrop-shaped areas of diseased plants develop downwind from the first infected plant.

Lesions appear on the tuber surface as irregular, brownish purple areas that have distinct, black, "eyebrow-pencil" marks at their margins. A brown, granular rot that lacks distinct margins extends a short distance into tuber tissue. If tuber lesions leak in storage, conditions favoring secondary bacterial invasion are created, and pockets of bacterial soft rot develop. If free moisture forms in storage, blight can spread to other tubers; little or no spread occurs if proper storage conditions are maintained.

Seasonal Development

The fungus that causes late blight, *Phytophthora infestans*, overwinters in infected tubers in storage, fields, or cull piles. Development and spread of the disease occur during periods of sprinkler irrigation, rainfall, or high humidity (above 90%) when the average temperature is below 78°F (25°C). When weather conditions are favorable, spores from infected foliage, cull piles, or infected tubers close to the soil surface move in air or water and spread the disease to other plants. Potato foliage and stems become infected at the soil surface when spores from infected tubers move up through the soil. Infected volunteers in fields or cull piles are usually the most important source of late blight inoculum, but in some cases the source is not known. Little inoculum is required for late blight to become serious; a few spores from one source can devastate an entire field if conditions are favorable for the disease. Tubers at or near the soil sur-face are infected by spores washed into the soil from leaves. Tubers may also be infected if they come in contact with spores during harvest.

Management Guidelines

Plant certified seed tubers, remove cull piles or treat plants that emerge from them with an appropriate herbicide, and control potato volunteers to minimize sources of late blight. Prevent tuber infections by using hilling operations that will keep tubers covered with soil until harvest. Leave tubers in the ground for at least 2 weeks after vine death to allow *Phytophthora* spores produced on the foliage to die before tubers are harvested; otherwise, even a small amount of foliar blight can cause substantial tuber decay in storage. Avoid harvesting under wet conditions; if wet tubers go into storage, use increased ventilation to remove free moisture from them as quickly as possible. Keep the storage temperature as low as allowed by the destined market and maintain proper ventilation to avoid free moisture in the piles.

In the coastal valleys of Washington, Oregon, and California, where late blight occurs every year, fungicide applications are recommended. Foliar-applied protectant and systemic fungicides are available. Protectant fungicides must be applied before blight lesions develop if they are to be effective. Systemics require less frequent application and will reduce disease if applied after lesions begin to develop, because they are translocated within the plant; however, they are more effective when applied before lesion development. Strains of the fungus have developed that are resistant to systemics; therefore labels require that they be applied in combination with protectant fungicides. Begin fungicide applications when weather favorable for blight occurs (rainfall or high humidity and average temperatures below 78°F [25°C]), and continue applications as long as these weather conditions continue. Repeat applications of foliar fungicides every 7 to 10 days. Systemics usually are applied only a few times each season; the frequency depends on local conditions. Applications of systemics must be followed by rainfall or irrigation within 24 hours to wash the material into the soil; most efficient uptake of the translocated material occurs through the roots.

In areas where late blight occurs sporadically, apply fungicides as soon as lesions appear and field conditions favorable for blight development are expected to continue for more than 2 or 3 days.

Late Blight Forecasting

Successful programs have been developed in the eastern United States for scheduling fungicide applications to control late blight. Daily records of temperature, rainfall, and relative humidity are used to forecast the development of late blight. Late blight forecasting programs are being studied to see if they can be adapted to growing areas in the western states, but as yet no recommendations are available for their use.

White Mold
Sclerotinia sclerotiorum

The fungus *Sclerotinia sclerotiorum* infects a large number of crop and wild plant species. In potato-growing areas it commonly causes disease on beans, lettuce, carrots, and alfalfa. Its occurrence on potatoes has increased with the use of sprinkler irrigation. White mold on potatoes is seen regularly in areas of Idaho, Washington, Oregon, and California, and is controlled by managing irrigation and avoiding excess nitrogen fertilization.

Sclerotinia overwinters as hard, black, irregularly shaped sclerotia that are formed inside dying potato stems. When exposed to moisture for prolonged periods above 60°F (16°C), sclerotia germinate and grow into light brown, funnel-shaped or flat fruiting structures called apothecia, which are 1/4 to 3/8 inch (6 to 9 mm) in diameter. These structures eject spores that infect nearby plants or can be carried for miles on wind currents. Most spread occurs within the potato field. The spores germinate and infect leaves or stems when free moisture is present for at least 48 hours. Water-soaked lesions and white mold develop on leaves and stems, and stems may be girdled by lesions. Tubers near the soil surface can be infected, but this phase of the disease is seldom seen in the western states.

White mold is associated with conditions favoring prolonged presence of water on stems and leaves. It is more prevalent in the wettest parts of fields and tends to occur after the vines close, when free moisture remains longer on stems and leaves. White mold can be severe if it occurs early in the season but can generally be controlled with proper irrigation management.

Avoid excess nitrogen, which gives heavy canopy growth that promotes conditions favorable for white mold. Watch for the appearance of apothecia or disease symptoms after plants have emerged and before vines close; if white mold infections appear, reduce irrigation to allow lower parts of the plants to dry out. After vine closure, apply water less often to make sure that lower leaves and stems do not remain wet continuously for periods of 48 hours or longer. Reduce irrigation rates if white mold appears on fallen leaves. Be sure changes in irrigation schedules do not allow soil moisture to drop below adequate levels.

Powdery Mildew
Erisyphe cichoracearum

Powdery mildew is known to occur on potatoes in the Snake River Valley, the Columbia Basin, and in Utah. It requires control only in furrow-irrigated fields in some Columbia Basin areas.

Powdery mildew first appears as very small, elongated, light brown stipples on stems and petioles. These coalesce as they develop to form dark, water-soaked lesions; symptoms spread to leaves. Sometimes the spore-forming structures of the fungus develop in leaf lesions, giving leaves a powdery, dusty brown or gray color that looks like soil or spray residue. As the disease progresses, lower leaves turn yellow and fall off. The plant stays erect but becomes covered with powdery mildew.

Sulfur will control powdery mildew if it is applied before infections are established. Symptoms appear after infections are established; sulfur applications made after this time are ineffective. In Columbia Basin furrow-irrigated fields where the disease is expected, first sulfur applications are made usually in mid to late July and repeated every two weeks. Norgold Russet is highly susceptible to powdery mildew and requires more careful controls when grown in fields where powdery mildew occurs. Consult your local extension agent, farm advisor, or other experts for specific recommendations in your area.

VIRUSES AND VIRUSLIKE DISEASES

The causal agents of virus and viruslike diseases are viruses and mycoplasmas. Viruses cannot reproduce outside a host organism and are transmitted mechanically or by vectors. Mycoplasmas are bacterialike microorganisms that are transmitted by aphids or leafhoppers that feed on potato plants. Certified seed programs provide the most effective means for controlling these diseases, because in most instances the primary source of inoculum is infected seed tubers. Insecticides are used to control the primary aphid vectors of potato viruses and reduce the spread of these diseases.

Leafroll
Potato Leafroll Virus

In the past, leafroll has caused severe losses in yield and value of stored tubers, especially of Russet Burbank. Aphid monitoring programs and control of aphid weed hosts, treatment of aphid overwintering hosts, elimina-

Light brown lesions and water-soaked areas with white, fluffy mycelium develop on stems infected by *Sclerotinia*.

The earliest symptoms of powdery mildew are small, light brown stipples on leaflets, stems, and petioles.

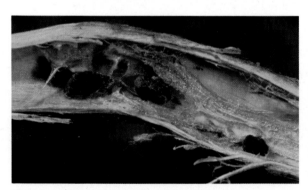

Sclerotinia forms hard, black sclerotia inside dead stems.

When spore-forming structures develop, powdery mildew lesions appear dusty brown or gray and powdery.

tion of potato volunteers, use of certified seed tubers, and increased use of systemic insecticides have reduced the incidence of leafroll to low levels in recent years. However, the disease can still be a problem when control measures are ignored.

Symptoms and Damage

The greatest losses to leafroll are from rejection of seed lots and devaluation of tubers with a speckling or netting of discolored tissue called net necrosis. Yield losses due to leafroll virus infection may also be significant. Russet

Burbank is the main cultivar that develops net necrosis, while some cultivars do not express this symptom under any condition. Leafroll virus symptoms vary according to time and method of infection.

Current-season infections occur when plants are infected with leafroll by aphid vectors. Symptoms develop first on young leaves at the tops of plants because the virus tends to be active only in rapidly growing tissue. Young leaves stand upright, roll upward, and turn pale or yellow. A pink or reddish color develops on the leaves of red-skinned cultivars, starting at the margins; the leaves of white-skinned or russet cultivars often

Upper leaves of plants with current-season leafroll virus infections stand upright, roll upward, and turn pale.

A pink color develops on the leaf margins of red-skinned varieties infected with potato leafroll virus.

Plants with chronic (tuberborne) leafroll virus infections are stunted, and symptoms develop on the lower leaves.

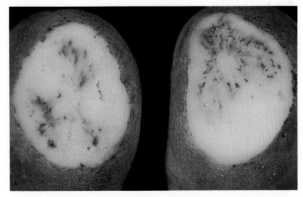

Brown discoloration of tuber vascular tissue, called net necrosis, may develop in most of the tubers from a plant infected with potato leafroll virus.

turn yellow. Maximum symptom expression occurs at 65° to 75°F (18° to 24°C). When primary infections occur late in the season, after new vine growth has stopped, foliar symptoms may not develop. Symptoms of blackleg, aster yellows, and witches' broom can be confused with current-season leafroll symptoms. Injury from drift of herbicides or pesticides may also be confused with leafroll. Identification of potato leafroll virus using serological techniques or experimental transmission to indicator plants is the only certain means of diagnosis. Current-season infections rarely cause economic yield losses in commercial potatoes except in cultivars that develop net necrosis.

Chronic or seedborne infections occur when plants grow from seed tubers infected with leafroll virus. Symptoms of chronic infections develop on lower leaves, which roll up, turn slightly pale, become stiff, dry, and leathery, and make a papery sound when rubbed. Plants are often stunted or rigid with leaf yellowing, and the older leaves may become discolored or turn brown and die. Chronic infections are much more severe than current season infections, but their incidence is greatly reduced by the use of certified seed tubers.

In certain varieties, including Russet Burbank and Targhee, the food-conducting tissue (phloem) in stems, petioles, and tubers may develop a discoloration in response to current season leafroll virus infection. In tubers, a translucent to dark brown speckling or netting called *net necrosis* develops; it may appear in the field or in storage. Not all tubers from an infected plant of a susceptible cultivar develop net necrosis. The actual number depends on the timing of virus infection and is greatest when infection occurs at the time of most rapid tuber expansion.

Other diseases or disorders cause tuber symptoms that may be confused with net necrosis. These include aster yellows, internal heat necrosis, internal brown spot, and discolorations of stem-end vascular tissue caused by Verticillium wilt, or vine killing under certain conditions. When potato leafroll virus is suspected, an experienced plant pathologist should be consulted to assure proper diagnosis.

Seasonal Development

Potato leafroll virus (PLRV) is carried in infected tubers and potato family weed hosts, such as nightshades and groundcherries, and is transmitted from plant to plant by specific aphid vectors. The green peach aphid, *Myzus persicae*, reportedly transmits potato leafroll virus most efficiently. This aphid acquires the virus when feeding on infected host plants, and can infect other plants after a latent period of about 24 hours or longer. Once infected with leafroll virus the aphid usually remains infectious for the rest of its life. Most transmission is by winged forms of the aphid over short distances within a field. The virus can also be transmitted by wingless forms to adjacent plants, usually within a row, or over long distances when winged forms are carried by the wind.

Most potato family plants can be hosts for potato leafroll virus. However, the most important virus sources are plants grown from infected seed tubers and volunteers that have grown from infected tubers left in the field or in cull piles.

Potato leafroll virus is introduced into the phloem, or food conducting tissue, of plants when virus-infected aphids feed. It moves through the phloem to growing leaves and expanding tubers; the more rapidly tubers are expanding at the time of infection, the more likely they are to be infected by the virus. If plants are infected when foliage growth has stopped but tubers are expanding, leafroll virus will infect the tubers without causing foliar symptoms.

If a cultivar is susceptible to net necrosis, most of the tubers infected with potato leafroll virus will develop this symptom. It develops first in the stem end and spreads further into the tuber in the field or in storage. Infected tubers may be symptomless at harvest and still develop symptoms during storage.

Management Guidelines

The principal measures for controlling leafroll are the use of certified seed tubers, insecticides to control aphid vectors, and control of potato volunteers. Some varieties, such as Centennial, are less susceptible to infection by the virus. Most varieties are resistant to the development of net necrosis; Russet Burbank is the most susceptible.

Grow seed potatoes in areas where sources of potato leafroll virus are limited and where aphid vectors do not appear until late in the season. If early season leafroll symptoms appear, remove each plant with symptoms and adjacent plants according to the diagram shown in Figure 30. Kill vines as early as possible; aphid trapping records can be used to time vinekill so that vines are dead before the vector population in the area builds up.

Minimize leafroll in commercial fields by planting certified seed, following recommended spray programs for controlling aphid vectors (see discussion of green peach aphid), and controlling volunteers. Vector control is important early and in midseason. Late-season control is important where cultivars susceptible to net necrosis are grown. Late-season infections result in little or no leafroll but will cause net necrosis symptoms in susceptible cultivars such as Russet Burbank.

When certified seed is used, volunteers become the major source of potato leafroll virus; use foliar sprout inhibitors to reduce volunteers. Always leave the field as clean as possible after harvest. In winter growing areas, disc the field immediately after harvest to expose tubers to as much heat and dessication as possible. In areas with cold winters, leave tubers exposed until after the first killing frost, then disc them under. If excessive numbers of tubers have been left in the field, you may want to use rototilling or some other type of additional tillage to destroy as many of them as possible. Do not leave cull piles in the field or around storage or packing facilities. Whenever possible, destroy volunteers that do appear in or near potato fields, as well as weeds such as nightshades, groundcherries, and mustard family weeds, that host the virus or aphid vectors. Use appropriate herbicides or cultivation to control volunteers that appear in rotation crops or field borders.

Net necrosis symptoms will intensify in storage; therefore, market tubers from fields with leafroll symptoms early and save tubers from fields without leafroll symptoms for prolonged storage. Storage at lower temperatures will retard but not prevent the development of net necrosis.

Mosaic Type Potato Viruses

The Mosaic-type potato viruses occur wherever potatoes are grown; potato viruses A (PVA), S (PVS), X (PVX), and Y (PVY) are of concern in the western states. They are called latent viruses because they can be carried in plants or tubers and may reduce yields without causing symptoms. Each virus has a number of different types or strains, and the reaction caused by these strains may differ. Potato variety and weather conditions also affect the symptom expression; the most pronounced symptoms appear in cool, cloudy weather. The presence of more than one of the viruses in a plant usually affects the types of symptoms and increases symptom severity. Symptoms caused by different viruses can be similar, so the type of virus usually cannot be identified by symptoms alone. Field diagnosis is often limited to "mosaic virus." Positive identification requires the use of indicator plants or serological techniques.

All of these viruses can be carried in tubers that show no symptoms. Some of them can be transmitted mechanically by seed cutting or by contact of a cut or open surface of an uninfected plant with the sap of an infected plant. Mechanical spread occurs more readily between plants when they are wet, and can be caused by cultivation or other movement of equipment, people, or animals through potato fields. Potato viruses A and Y are also transmitted by aphids. The most important aphid vectors are the green peach aphid and the potato aphid. Aphids can transmit either of these viruses immediately after feeding on an infected plant, but lose the virus after they feed on one to three uninfected plants. These styletborne viruses are usually transmitted only short distances, and because aphids are infective for only short times, controlling aphid populations is not the most effective way to control spread of these viruses. Potato viruses S and X are usually not spread by aphids.

The major control for potato viruses is the use of virus-free seed tubers. Special procedures are used to obtain virus-free plants and to identify the presence of these viruses including potato leafroll virus in tubers to be used for seed production. Steps must be taken to prevent virus infection during the production of certified seed tubers (see *Seed Quality and Seed Certification*, in the chapter, *Managing Pests of Potatoes*). In all potato production areas, disinfect seed cutting equipment with recommended disinfectants (Table 7), avoid cutting seed tubers that have sprouts large enough to be broken, and control potato volunteers and weeds, especially nightshades, which can be hosts for potato viruses.

Potato Virus A (PVA)

The disease caused by potato virus A is known as *mild mosaic, crinkle mosaic,* and *veinal mosaic.* Leaves on affected plants are mottled, with some areas light green to yellow and some darker green than normal. Mottled areas may vary in size and occur both on and between leaf veins. Margins of affected leaves may be wavy, and leaves may appear slightly rugose where veins are sunken and interveinal areas raised. Rugose symptoms can also be caused by certain strains of potato virus X and potato virus Y. Affected plants tend to open up because the stems bend outward. Symptoms are more pronounced and yield reductions are greater when potato virus X is also present.

Potato virus A is styletborne by at least seven different aphid species, including the green peach aphid, *Myzus persicae,* and the potato aphid, *Macrosiphum euphorbiae.*

Potato Virus S (PVS)

Although in most cultivars potato virus S is symptomless, some potato cultivars infected with some strains of the virus demonstrate slight deepening of leaf veins, and possibly stunting, mottling, or bronzing. There is argument about whether this virus causes any reductions in yield by itself, although losses of 10% to 20% have been reported. Combined presence of potato viruses S and X reduces yields more than either virus alone.

Mature plants are resistant to the virus; plants must be infected early in the season for tubers to become infected. The disease is carried in tubers and is mechanically transmitted. Potato virus S is controlled by using certified seed tubers. Special laboratory or greenhouse techniques must be used to identify the presence of potato virus S in seed stock because it is almost always latent.

Potato Virus X (PVX)

The most widespread of the potato viruses, potato virus X is also called *potato latent virus, potato mottle virus,* and *latent mosaic.* Some strains of the virus produce no visible symptoms when no other viruses are present, although symptomless plants may have yields reduced 15% or more when compared to virus-free plants. Other strains of potato virus X may cause mild mottling, sometimes called weather mottling, under prolonged periods of low light intensity. Low temperature (60° to 68°F [15° to 20°C]) enhances symptoms, and the additional presence of potato virus A or Y may cause crinkling, rugose mosaic or browning of leaf tissue. Some combinations of virus strain and potato variety may result in death of all or part of the plant and tuber necrosis.

Potato virus X is carried in tubers and can be transmitted mechanically by machinery, spray equipment, root-to-root contact, sprout-to-sprout contact, or seed cutting equipment. There are no known aphid vectors. The spread of potato virus X is controlled by using certified seed tubers and avoiding mechanical transmission.

Potato Virus Y (PVY)

The most severe of the potato viruses, potato virus Y causes symptoms known as *severe mosaic, leafdrop streak,* and, when potato virus X is also present, *rugose mosaic.* Potato virus Y is sometimes called *potato vein banding virus.*

Two types of symptoms are characteristic of potato virus Y alone. *Vein banding* is the development of brown

Leaflets of plants infected with mosaic virus (*on the right*) may be mottled with areas lighter green than normal and areas darker green than normal. Leaflets become wrinkled because the darker areas grow faster. Healthy leaves are shown on the left.

Potato virus Y causes vein banding, the development of brown to black streaks along veins on the undersides of leaves.

Rugose mosaic symptoms (*left*), where veins appear sunken and interveinal areas raised, occur on leaves of plants infected by both potato viruses X and Y. Certain strains of X or Y alone may cause rugose mosaic. Healthy leaves are shown on the right.

streaks along veins on the undersides of leaves. If vein banding becomes severe, *leafdrop streak* results. Lower leaves die, sometimes leaving a tuft of wrinkled leaves at the top of the plant with some of the dead leaves clinging to the stem, resulting in a palm tree appearance. Vein banding and leafdrop streak are usually more severe in current-season infections. Plants that grow from infected seed tubers may be dwarfed and may have mottled, wrinkled, and brittle leaves. They develop milder vein banding symptoms.

Plants that are infected with potato virus Y may develop a mosaic mottling that is similar to symptoms caused by potato virus A, but usually appears as smaller, more numerous discolored areas. Mottling may be masked at temperatures below 50°F (10°C) or above 70°F (21°C). When potato viruses Y and X are present in the same plant, they may cause a severe wrinkling of the leaves known as rugose mosaic; veins are sunken and interveinal areas are raised. Milder rugose symptoms may be caused by strains of potato virus Y or X alone.

Potato virus Y has a wide host range that includes tomatoes, peppers, eggplants, legumes, nightshades, pigweeds, and many other weed species. The only important overwintering source of potato virus Y is infected potato tubers. The virus is transmitted by at least 25 different aphids, the green peach aphid being the most important. For control of potato virus Y, plant certified seed tubers, eliminate cull piles, control potato volunteers and weeds, and prevent buildup of aphid populations.

Corky Ringspot
Tobacco Rattle Virus (TRV)

Tobacco rattle virus, transmitted by stubby-root nematodes (*Paratrichodorus* spp.), is the cause of corky ringspot in potato tubers. Corky ringspot symptoms are sometimes called *spraing* or *sprain*. The disease occurs in certain areas of California, Colorado, Oregon, Idaho, and Washington, and is controlled by planting virus-free seed tubers, eliminating weed hosts of the virus, controlling stubby-root nematode, and avoiding fields with a history of corky ringspot.

Corky ringspot symptoms in tubers vary depending on virus strain, potato variety, temperature, and time of infection. Two types of symptoms are most common. The first is visible as rings of dark brown, corky layers that alternate with rings of healthy tissue. These rings originate at surface lesions caused by nematode feeding. The second type of symptom is characterized by small, brown flecks that are diffused through the tuber. This second type of symptom may be confused with symptoms caused by certain strains of alfalfa mosaic virus, which are more common in some growing areas than corky ringspot. In White Rose and Kennebec, corky rings 1/4 to 1/2 inch (6 to 12 mm) in diameter appear in mature tubers; in Russet Burbank tubers, symptoms are spots, arcs, or rings of brown tissue. In some areas corky ringspot causes the greatest losses in processing potatoes, because processors usually will not accept tubers from fields with a known history of the disease.

Plants can become systemically infected with tobacco rattle virus. Aboveground symptoms are caused by certain strains of the virus and occur rarely, when plants grow from infected tubers. Some sprouts from a seed piece form stunted stems with leaves showing patterns of yellow lines, while other sprouts from the same seed piece grow normally.

Corky ringspot occurs on potatoes grown in coarse, sandy soils where the stubby-root nematode is present. The nematode acquires tobacco rattle virus by feeding on the roots of infected plants and spreads the virus to other plants as it feeds. Adult nematodes that acquire tobacco rattle virus remain infective for life, which can be as long as 2 years. A large number of weeds are hosts for tobacco rattle virus, including nightshades, pigweeds, shepherds-purse, purslane, common cocklebur, birdsrape mustard, and sunflower. Many crop plants can be infected by the virus without showing disease symptoms. Young potato roots and tubers are infected with tobacco rattle virus when virus-infected nematodes feed on them. The virus is also transmitted in infected tubers and the seeds of hairy nightshade, purslane, and common cocklebur; any of these can introduce corky ringspot to a new area, where it will become established if the stubby-root nematode is present.

For control of corky ringspot, plant certified seed tubers, eliminate weed hosts, and control stubby-root nematodes (page 115). Crop rotations are ineffective for control of this disease because both the virus and nematode have broad host ranges. When growing processing potatoes or seed potatoes, avoid fields with a history of corky ringspot. Most seed certification programs have a zero tolerance for tobacco rattle virus. In Idaho, seed potatoes cannot be grown for certification in a field that has a history of corky ringspot.

Calico
Alfalfa Mosaic Virus

Alfalfa mosaic virus causes the striking yellow symptoms on leaves known as calico that can be seen wherever potatoes are grown. Usually only an occasional plant is affected and the disease is unimportant, but it can become a problem when potatoes are planted next to alfalfa or clover, the normal hosts for this virus. Some strains of alfalfa mosaic virus cause cell death, or necrosis, rather than the pale to bright yellow mottling or blotching of leaf surfaces. Necrosis occurs mainly in the stems and can move into tubers, where it appears throughout as dry,

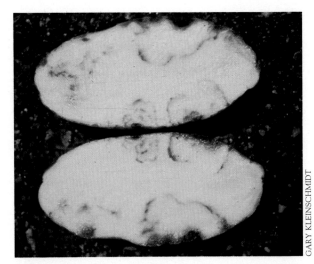

Corky ringspot in Russet Burbank tubers may appear as arcs or concentric rings of brown tissue.

Alfalfa mosaic virus causes a pale or bright yellow mottling of leaves, called calico.

corky areas or rusty brown patches. Necrosis may also occur in tubers of plants with characteristic calico symptoms. Tuber necrosis caused by alfalfa mosaic virus may be confused with corky ringspot; however, plants that have tubers with corky ringspot symptoms rarely show any aboveground symptoms.

Alfalfa mosaic virus is styletborne by at least 16 different aphid species, including the green peach aphid. Transmission of the virus is most likely to occur when aphids migrate into potato fields after an alfalfa field has been mowed. The pea aphid, *Acyrthosiphon pisum*, is probably the most important vector; it is most likely to migrate from mowed alfalfa into potato fields. Aphids do not transmit the virus between potato plants, but it can be carried in infected tubers. For control of alfalfa mosaic, plant certified seed tubers and avoid planting near alfalfa or clover, especially downwind. Older alfalfa fields are much more likely to be infectious because of virus buildup. When growing seed potatoes, remove calico plants as soon as they appear.

Curly Top
Sugarbeet Curly Top Virus

The beet leafhopper, *Circulifer tenellus*, transmits curly top virus in a persistent manner. Curly top infects a wide range of other crop plants besides potatoes, including tomatoes, peppers, sugar beets, beans, and melons. Both the virus and the leafhopper survive on a large number of wild plants. Curly top is considered of little importance in most potato-growing areas, but it has been common in southwestern Idaho, California's Kern County, New Mexico, and Utah.

Curly top symptoms include dwarfing, yellowing, and rolling of upper leaves. Leaves near the growing point develop yellow margins and become elongated, cupped, twisted, and rough. Outer leaflets are rounded at their tips; their veins remain green while the rest of the leaflet turns pale yellow. Aerial tubers may form. The virus will infect tubers, and if they are planted they will fail to emerge or will develop stiff, erect stems that are pinched together at their tips and have leaflets that do not unfold completely.

No specific controls are used for this disease in potatoes. In areas where curly top is a problem in other crops, spray programs to control beet leafhoppers in native vegetation have reduced the incidence of curly top in potatoes.

Aster Yellows

Aster yellows is caused by a mycoplasma, an organism similar to bacteria, that infects a wide range of plants. The disease is also known as *potato late-breaking virus*, *purple top*, and *purple top wilt*, and is of minor importance in potatoes in the western states. It is transmitted to potatoes from other host plants by the aster leafhopper, *Macrosteles fascifrons*, and may occur in desert areas because of the presence of Russian thistle, a favorite host of the leafhopper and a reservoir for the mycoplasma. Aster yellows mycoplasma is carried in infected tubers, but there is no spread between potato plants by the leafhopper.

Symptoms of aster yellows include an upward curling and yellowing or purpling of top leaves that resembles leafroll, except that the discoloration is more intense. Top leaves may be very light green with deep veins, and plants may be stunted. Aerial tubers usually form in the axils of stems. Scattered brown spots of dying tissue may develop throughout infected tubers, a symptom that may be confused with leafroll net necrosis. Tuber necrosis caused by aster yellows mycoplasma is scattered throughout the tuber, rather than being more concentrated in the vascular ring and at the stem end of the tuber as is leafroll net necrosis. Aster yellows is considered self-eliminating because infected tubers either fail to produce plants or they produce plants that are severely stunted and fail to form tubers. No control measures are used for this disease in potatoes; some seed certification programs have tolerance levels for aster yellows.

Witches' Broom

Witches' broom is caused by a mycoplasma that is spread to potato plants from other host plants by leafhoppers of the genus *Scleroracus*. Leaves on infected plants may curl and discolor similarly to those affected by leafroll. The mycoplasma survives in tubers, and when infected seed tubers are planted they develop numerous stems with small, round leaves and form large numbers of pea-sized tubers. Because no useful tubers are produced on plants grown from infected seed tubers, witches' broom like aster yellows is self-eliminating. This disease occurs rarely, and no specific controls are recommended.

Physiological Disorders

Environmental factors including improper cultural practices, high or low temperatures, and inadequate, excess, or uneven levels of soil moisture or nutrients may disrupt normal tuber growth and development and cause undesirable disorders. Disorders not caused by biological agents (microorganisms or insects) are called *physiological* or *abiotic disorders*.

Physiological disorders can affect the internal as well as external appearance of tubers. Tubers used for seed, fresh market, or processing may be affected; certain disorders are more important if tubers are intended for one of these markets than if they are intended for the others. Losses from most physiological disorders can be avoided or minimized by using good cultural practices, particularly those that provide uniform and adequate soil moisture and fertility throughout tuber initiation and growth. However, the causes of some disorders are unknown.

Secondary Growth

When environmental stresses have temporarily halted tuber growth, resumption of growth may cause nonuniform tuber development called secondary growth. Hail or frost injury to leaves, as well as high soil temperature (above 81°F [27°C]), low moisture, an imbalance in fertility, or combinations of these factors can interrupt normal tuber growth. Potato cultivars that produce long tubers tend to be more susceptible to secondary growth, but no cultivars are completely resistant.

Because periods of high temperature and injury by hail or frost contribute to secondary growth disorders, these disorders cannot be avoided entirely. However, attention to good cultural practices can minimize secondary growth. Plant for uniform stands with correct spacings and numbers of stems per hill to ensure full tuber set. Maintain levels of soil fertility and moisture that support uniform plant development throughout the growing period to maintain uniform tuber growth. Irrigation to maintain recommended soil moisture levels during hot weather is critical, but excessive soil moisture must be avoided. Avoid excess nitrogen, which causes excessive vine growth.

Knobby tubers are formed when secondary growth occurs at lateral buds.

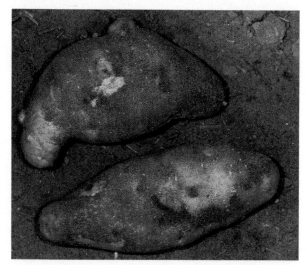

Pointed end, which is a constriction of either the stem or bud end of a tuber, results when secondary growth occurs either early or late in the season.

Growth cracks are splits in the tuber surface caused by rapid expansion of tuber tissue. They heal over but make the tuber unacceptable for fresh market.

Tuber Malformations. Several different types of tuber malformation can occur when growth resumes. Each type has a descriptive name. *Knobby tubers* are formed when secondary growth occurs at one or more lateral buds. Growth in a longitudinal direction results in *bottleneck*, *dumbbell*, or *pointed end*; the type of symptom is determined by the time at which the growth interruption occurs. Growth interruption during early tuber development results in bottleneck or pointed stem ends, mid-season interruption results in dumbbell, and late-season interruption tends to cause tubers with pointed bud ends. The constricted stem ends of pointed-end tubers often have abnormally high levels of reducing sugars, and develop a symptom called *translucent end*.

Heat Sprout. Secondary growth may occur as sprouts that can develop from buds on stolons or tubers. The secondary sprouts elongate and may remain underground or may emerge and develop into leafy stems. Heat sprout often occurs when growth resumes after being temporarily interrupted by high soil temperatures.

Tuber Chaining. Secondary growth may occur from a primary tuber as a series of tubers along a stolon. Tuber chaining frequently occurs after high soil temperature.

Little Tuber. Another form of secondary growth is little tuber, also called *no top*. This disorder involves the formation of many secondary daughter tubers from a planted seed tuber without foliage development. The daughter tubers may develop directly from buds, or may be formed on short stolons that emerge from buds on seed tubers. Little tuber is associated with improper storage handling of seed tubers. The disorder occurs most frequently when physiologically old seed tubers are planted, when seed tubers are planted into cold soil (below 50°F, 10°C) after they have been stored at a temperature of 68°F (20°C) or higher, or when seed tubers are placed in cold storage after sprouting and then planted at a later date. To control little tuber, store seed at 40°F (4.4°C), avoid using seed tubers that have been stored a long time, and do not plant older seed tubers in cold, dry soil.

Translucent End
Jelly End

The disorder translucent end, which is also called *sugar end*, *glassy end*, or *incipient jelly end*, occurs when the stem end of a tuber develops high reducing sugar levels. Usually less than 2 inches (5 cm) of the stem end is affected. The tuber tissue appears water-soaked or translucent. Tubers with translucent end produce french fries or chips that are undesirably dark in color because processing caramelizes reducing sugars. Translucent end tissue frequently becomes soft and jelly-like, a symptom called jelly end. Jelly end may develop in the knobs or pointed ends of malformed tubers. Under dry conditions, jelly-end tissue may become a crumbly, rubberlike mass, a condition called *jelly end rot*.

Translucent end and jelly end are associated with tubers showing secondary growth symptoms. Starch levels decrease in the stem ends of tubers undergoing second growth, and reducing sugar levels increase. Use the same cultural practices that reduce secondary growth to avoid translucent end and jelly end.

Growth Cracks

Sudden, rapid increases in volume of internal tuber tissue cause growth cracks, splits in the skin of tubers. The splits usually run lengthwise, and exposed tuber tissue heals, leaving fissures in the tuber surface. Tubers with external growth cracks are usually unacceptable for fresh market.

Irregular moisture conditions are usually responsible for growth cracks. A heavy rain or irrigation following a dry period will cause a rapid increase in growth activity. The resulting internal pressure may split the skin. Fertilizer applications that cause rapid tuber growth may also cause growth cracks. When conditions favoring growth cracks occur, the incidence of growth cracks is greater where plant spacings are wide or if fewer tubers are set per hill.

To reduce or eliminate growth cracks, provide uniform growing conditions by maintaining adequate soil moisture and fertility throughout the growing season. Use correct seed piece size and spacing to obtain adequate stands and the correct number of stems per hill.

Surface Abrasions

Tubers with surface abrasions have a feathery, ragged appearance, symptoms referred to as *skinning* or *feathering*. Most abrasions usually occur at the bud end of the tuber. Under proper wound-healing conditions, surface abrasions will heal. However, if exposed to high temperature and dry air, a scald symptom may develop, and the injured areas may turn dark and become susceptible to rot.

Surface abrasions are caused by rough handling of immature tubers. If tubers are harvested before their skins have developed fully, they are more susceptible to abrasion. Allow tubers to mature properly by letting soil fertility decrease before vinekill, and waiting 2 to 3 weeks between vinekill and harvest. Use careful harvesting and handling procedures, and always cover tubers during transit to prevent wind and sun damage.

Bruising

Three different types of bruising occur. *Blackspot bruise* and *shatter bruise* are caused by the bouncing of tubers against hard surfaces during harvesting and handling. *Pressure bruise* results from the pressure of other tubers or container surfaces during storage when humidity drops too low. Bruise damage is reduced by following careful harvesting and storage procedures.

Blackspot Bruise. Blackspot bruise appears as small, dark, discolored areas beneath the potato skin. Damage usually is not visible unless tubers are peeled. Tubers are initially bruised when they fall or are bounced around during harvesting and handling. Injured tissue first turns pink, then red, then darkens. When the temperature is about 70°F (about 20°C), discoloration begins about 6 hours after injury and is complete within 24 hours; discoloration occurs sooner at higher temperatures. Stem ends of tubers and tubers with lower water content are more susceptible to blackspot bruise. Inadequate potassium levels increase the potential for blackspot bruise. Cultivars vary greatly in their susceptibility.

Shatter Bruise. Impact of sufficient strength to rupture the tuber skin and several layers of cells beneath it causes shatter bruise. The resulting small splits in the skin may not be visible at first but are easy to see after the injured tuber tissue dries out. The margins of shatter bruises show the same discoloration as blackspot bruises, but the injuries are visible on the tuber surface and usually penetrate deeper than blackspot bruises. Tubers are susceptible to shatter bruise when their temperature is low, 45° to 50°F (7° to 10°C); tubers with higher water content are more susceptible.

Controlling Blackspot and Shatter Bruise. Blackspot bruise and shatter bruise are minimized by careful harvest of tubers that are not cold and that have the correct water content. To provide the best tuber hydration at harvest, you must follow good cultural and pest control practices to ensure adequate vine and root development. Make sure potassium nutrition is adequate by using a preplant soil test and applying the recommended amounts of potassium for your area. Allow tubers to mature after vines are killed. Maintain adequate soil moisture until harvest; avoid excess irrigation, which can cause decay, but keep soil moisture above 50% of field capacity. Avoid harvesting when tuber flesh temperatures are below 45°F (7.3°C). When night temperatures are below 50°F (10°C), harvest during the warmest part of the day. Use harvesting and handling procedures designed to reduce bruising; consult the References for specific recommendations to reduce bruising during harvest operations. You can use a bruise test to identify changes in operations that may increase or decrease bruising.

Bruise Tests. Blackspot and shatter bruises can be detected more quickly using tests. Heat or hot water may be used to hasten the development of bruise symptoms; chemical tests detect injured tissue before it develops bruise symptoms. These tests may be useful for determining where bruising injuries are occurring or whether modifications in harvesting or handling procedures have reduced bruising. Both of the chemical tests described below use toxic chemicals; put warning labels on all con-

Blackspot bruises are small, dark areas that develop just underneath the skin where a tuber has bumped against a hard surface.

Shatter bruises are small splits in the tuber skin. They are most visible after the injured tissue dries out.

tainers and always wear plastic gloves when performing the tests.

Heat. Both shatter and blackspot bruise can be detected more quickly by heating tubers. Hold tubers at 90° to 95°F (32° to 35°C) for 12 hours, then inspect for bruise symptoms.

Hot Water. The hot water test detects both shatter and blackspot bruise. Place tubers in hot water (140°F [60°C]) for 10 minutes. Remove them, let them stand for 6 to 7 hours, then inspect them for bruise symptoms.

Triphenyl Tetrazolium Test. Both shatter and blackspot bruise are detected by the triphenyl tetrazolium test.

1. Prepare a solution of 2,3,5-triphenyl tetrazolium hydrochloride in tap water by dissolving 1 1/2 ounces (40 g) of the chemical in 1 gallon of water.
2. Wash and peel the tubers to be tested.
3. Place peeled tubers in a 10-quart plastic pail and carefully pour the test solution over them. Allow them to stand in the solution for about 40 minutes.
4. Bruises begin showing up in 10 minutes as dark pink areas.
5. Remove tubers from the solution, check them for bruising, and destroy them. Store the test solution carefully, or dispose of it properly.

Catechol Test. Shatter bruise, but not blackspot bruise, is detected with the catechol test.

1. Prepare the test solution by dissolving 2 ounces (56 g) of technical grade catechol in 1 gallon of water in a plastic jug. Label the jug to clearly indicate that it contains a poisonous liquid.
2. Wash the tubers to be tested and place them in a 10-quart plastic pail.
3. Pour the test solution over the tubers. After a few minutes, remove the tubers and return the test solution to the plastic jug for future use.
4. After about 10 minutes, bruised and skinned areas develop a red color.
5. Destroy tubers after the test.

Pressure Bruise. Pressure bruises are flattened or sunken areas on the surfaces of tubers that result from low humidity in storage. When humidity is below 95%, pressure from storage surfaces or other tubers causes depressions in the tuber surface. Affected tissue loses water and the depressions persist if the humidity remains below 90%. Tuber quality is reduced and affected areas become very susceptible to blackspot bruise when handled. Some cultivars are more susceptible to pressure bruise; round white cultivars are more susceptible than

Russet Burbank. Keep humidity above 90% during storage to reduce or eliminate pressure bruising. Use reduced pile depths when storing susceptible cultivars.

Hollow Heart

The development of cavities in the pith area of tubers is known as hollow heart. The cavities may be very small or may take up the entire pith area, there may be one cavity or several, and cavities may develop near the stem end or near the bud end. In the case of *stem-end hollow heart*, an area of dead, brown cells develops before cavities form. This symptom is called *brown center* or *incipient hollow heart* and may develop while tubers are very small. Cells in the brown area later split apart and cavities form in the middle of the brown areas. *Bud-end hollow heart* usually appears late in the season and develops without any brown discoloration. The cavities are located nearer the bud end. A tan or brown layer of suberin that resembles tuber skin may form on the inside of hollow heart cavities. A type of black heart called "cat's eye," a physiological disorder with different causes, may resemble hollow heart; however, the small cavities are lined by a thick, gray or black layer.

Hollow heart is associated with periods of rapid tuber growth; potassium deficiency and stress that causes cell injury may contribute to the development of symptoms. Norgold and Lemhi usually develop bud-end hollow heart. Russet Burbank may develop brown center and stem-end hollow heart or bud-end hollow heart. Cool soil temperatures during tuber initiation induce brown center development; 5 days of 64°F (18°C) or lower day temperature and 50°F (10°C) or lower night temperature will induce brown center. More hollow heart develops in Russet Burbank tubers that are initiated in wet soil.

To decrease the incidence of hollow heart, use close plant spacings and large seed pieces that will ensure a good stand with an adequate number of stems per hill. This will provide full tuber set and reduce the likelihood of periods of rapid tuber growth. Keep soil moisture and fertility adequate to maintain uniform tuber growth rates. When tubers are forming, irrigate carefully during cool weather to avoid wet soil.

Internal Brown Spot

The disorder internal brown spot consists of small, round or irregularly shaped, light tan or reddish brown, corky spots that are scattered throughout the tuber tissue. This disorder is also known as *internal browning, internal rust spot, physiological internal necrosis, chocolate spot,* and

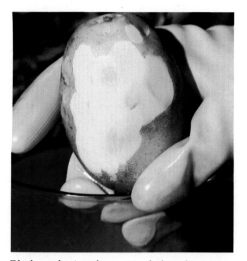

Blackspot bruises show up as dark pink areas about 10 minutes after immersion in a triphenyl tetrazolium test solution.

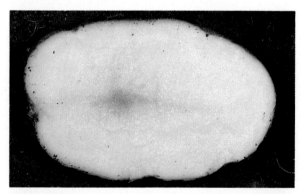

Brown center, or incipient hollow heart, is an area of dead, brown cells that develops in the center of tubers.

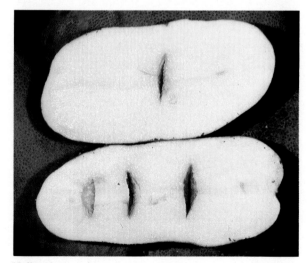

Hollow heart refers to a cavity that forms in the center of a tuber. It may or may not be surrounded by an area of dead, brown cells.

internal brown fleck. Internal brown spot lesions may occur anywhere within the tuber, but they seldom occur in the cortex, outside of the vascular ring. The lesions may develop anytime during the season or in storage.

The cause of internal brown spot is unknown. However, this disorder is associated with certain conditions: hot, dry weather, low soil moisture, sandy soils (compared to soils high in organic matter), uneven growth rates due to temperature or moisture conditions, and low calcium levels in tuber tissue. Some varieties are more susceptible to internal brown spot, and the disorder tends to occur more frequently in larger tubers. Tubers that are mature at harvest and tubers stored at lower temperatures develop less internal brown spot.

To reduce the occurrence of internal brown spot, provide proper soil fertility and moisture and maintain uniform growing conditions throughout the growing season. Allow tubers to mature between vinekill and harvest, but do not allow tubers to lie in hot, dry soil. Cut samples of tubers at harvest; store any lots that show internal brown spot symptoms as close to 40°F (4.4°C) as possible.

Internal Heat Necrosis

Symptoms of internal heat necrosis are similar to those of internal brown spot but occur mainly in the area of the vascular ring. Internal heat necrosis is caused by high soil temperatures between the time vines begin to die and harvest time. Tubers in the upper 2 inches (5 cm) of soil are most likely to develop symptoms, which appear at or near harvest. Use irrigations, mulching, or increased soil cover to reduce exposure of tubers to high temperatures. If you use irrigation during hot weather to cool the soil, be sure to keep the soil below field capacity. Wet soil during hot weather causes black heart or tuber rot. If possible, harvest early to avoid high soil temperatures.

Vascular Discoloration

Vine killing when soil moisture is below 50% of field capacity or vine-killing frosts that occur in dry weather may cause vascular discoloration in tubers. The discoloration occurs in the region of the vascular ring, and may be light tan, reddish brown, or dark brown. Vascular discoloration may be confined to an area near where the stolon attaches to the tuber, or may extend throughout the tuber. This disorder is also known as *stem-end discoloration, browning,* or *necrosis, vascular browning,* and *phloem necrosis.* Vascular discoloration may be present at harvest or develop during the first 2 months of storage.

Similar symptoms may be caused by leafroll virus or *Verticillium.* Leafroll virus causes net necrosis in the tubers of some cultivars, a discoloration that is restricted to phloem tissue, but extends throughout affected tubers. Tubers with leafroll net necrosis can be distinguished by laboratory detection of the virus or by planting the tubers, which grow into plants with chronic leafroll symptoms. Tubers on plants infected by *Verticillium* may develop discoloration of xylem tissue; however, it is confined to the stem end and is usually darker than physiological vascular discoloration. Vines usually show symptoms of Verticillium wilt before they are killed, distinguishing the biotic disease from the physiological disorder.

To control vascular discoloration, maintain adequate soil moisture during vine killing operations. Avoid killing stressed vines.

Black Heart

Black heart is a dark gray, purple, or black discoloration of the tuber pith area that is caused by insufficient oxygen. Brown, purple, or black patches may be visible on the tuber surface on rare occasions. Cavities resembling hollow heart occasionally develop, but unlike hollow heart they are surrounded by gray or black layers. These cavities are sometimes referred to as *"cat's eye."* Black heart may develop in the field or in storage where reduced gas exchange occurs.

Black heart results when internal tuber tissue does not receive enough oxygen to support respiration. In the field, black heart is usually caused by waterlogged soil or extremely high soil temperatures. Waterlogging prevents oxygen from getting to the tubers; high soil temperatures increase respiration rates so that oxygen cannot reach internal tuber tissues fast enough. In storage or transit, black heart results from poor ventilation, improper heating of tubers during shipment to prevent freezing, or prolonged storage at temperatures near freezing.

Potato cultivars differ in their susceptibility to black heart, but precautions must be taken with all of them. Make sure that soil drainage is good and that there are no spots in the field where water collects. Do not leave tubers in or on hot soil. Store tubers at the proper temperature with adequate ventilation. Avoid exposing tubers to temperatures above 90°F (32°C).

Low Temperature Injury

Temperatures below freezing may injure tubers. The severity of the damage depends on the temperature and duration of exposure. The flesh of tubers exposed to temperatures slightly below freezing may turn gray or reddish brown, a symptom called *mahogany browning*. Tubers exposed to temperatures below 28°F (−2°C) for extended periods sustain serious injury. If cells are not frozen, symptoms are not visible on the tuber surface, but internal cells die and tissues turn brown, causing the symptoms of *chilling injury*. If ice crystals form in the cells, *freezing injury* results. Freezing injury usually occurs on one end or side of the tuber that is closest to the soil surface, usually the bud end, and symptoms are easily seen on the tuber surface. Tissues at the stem end and in the vascular ring are more sensitive to low-temperature injury. Affected tissues become liquid, and the vascular ring may break down completely. Frozen tissue may dry out and become tough or chalky. Use proper hilling to help protect tubers from freezing temperatures. Whenever possible, harvest and store tubers before severe frost is likely. Use high airflow to reduce storage losses in lots that were exposed to freezing temperatures in the field.

Greening

Tubers that are exposed to sunlight in the field, in storage, or at market may develop greening. The green pigment chlorophyll is formed just beneath the skin and tubers develop a green color. This symptom may also be called *greened tubers*, *sun green*, *virescence*, or *sunscald*. The chlorophyll is harmless, but levels of other compounds, called glycoalkaloids, also increase in the tissue exposed to sunlight. They are mildly toxic and give tubers a bitter taste, making them unpalatable. Tubers with greening are unacceptable for fresh market or processing. White-skinned varieties such as White Rose and Kennebec are more susceptible to greening. Use careful hilling operations early in the season to make sure tubers remain covered with soil throughout the season. Certain cultivars such as Kennebec form tubers shallowly and require deeper planting or higher hills. Broad hills are better than narrow, peaked hills. Additional hilling after vine death may be necessary.

Enlarged Lenticels

Excessive moisture in contact with tuber surfaces can cause enlarged lenticels, also known as *water spots* or *water scab*. Tissue beneath lenticels (the small pores present in the skin of tubers) swells, bursts through the protective suberin layer beneath the lenticels, and forms raised masses of corky tissue on the tuber surface. Enlarged lenticels resemble scab lesions, but are smaller and lighter in color; they are susceptible to infection by decay organisms. Enlarged lenticels usually are caused by excessive soil moisture in the field, but may also form in storage if free moisture is allowed to stand on tuber surfaces. Avoid overwatering or ponding of water in the field, and use forced air ventilation to dry wet tubers as quickly as possible after harvest. Avoid condensation on tubers during storage by using proper ventilation, humidity, and temperature controls.

Elephant Hide

The development of thick, coarse russeting, primarily on Russet Burbank tubers, is called elephant hide or *alligator hide*. It is occasionally seen on other russet cultivars. Furrowing and cracking of a thickened periderm give affected tubers the appearance of an elephant's hide. A similar condition occurs on smooth-skinned cultivars, where it is called *fishy skin* or *turtle back*. The causes and control of this disorder are not known. It may be caused by environmental factors, contact with decaying organic matter or with salts in the soil, or certain soil-applied chemicals.

Aerial Tubers

Severe restriction of carbohydrate movement from vines into growing tubers causes the formation of small tubers at the base of stems or in leaf axils, where leaves join stems. These tubers are called aerial tubers, *stem tubers*, or *air tubers*. One or several tubers may form on a plant, usually close to the ground. Aerial tubers are green to purple in color, usually small, and oddly shaped. Aerial tubers may be caused by various diseases, including blackleg, Rhizoctonia, curly top virus, aster yellows, and witches' broom, by insect injury to stems or psyllid toxin, or by mechanical injury or waterlogging of the soil.

Ozone Injury

The air pollutant ozone frequently reaches levels in southern California potato-growing areas that damage sensitive cultivars. Ozone injury also occurs in the southern San Joaquin Valley. Symptoms of ozone injury, frequently called *speckle-leaf*, appear as tiny spots of dying tissue on upper leaf surfaces and bronzing on the lower surfaces. Yields may be reduced. If injury is severe, plants are defoliated. Avoid planting sensitive cultivars such as Centennial Russet and Red LaSoda in areas where ozone frequently reaches damaging levels. Damage is accentuated by low nitrogen.

ARTHUR KELMAN

Internal brown spot refers to small spots of light tan or reddish brown corky tissue that are scattered throughout the tuber tissue.

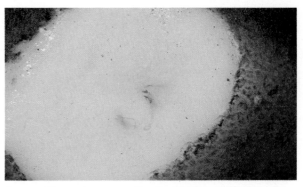

Stem-end browning or vascular discoloration is a light brown to dark brown discoloration of the vascular tissue at the stem end of tubers. It can be caused by the Verticillium wilt fungus or physiological stress during vinekill.

Elephant hide, or alligator hide, refers to the development of thick, furrowed, and cracked skin that occurs primarily on Russet Burbank tubers.

If tuber tissue is frozen, a dark line develops between injured and uninjured tissue. Affected tissue becomes water-soaked and is easily seen on the surface. Frozen tissue usually disintegrates.

Enlarged lenticels are raised masses of corky tissue that protrude through the lenticels of tubers exposed to excessive moisture.

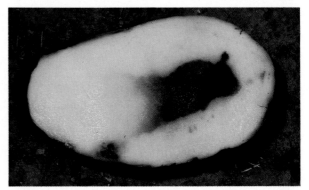

Blackheart is a dark gray, purple, or black discoloration in the center of a tuber exposed to waterlogged soil or high soil temperatures.

Herbicide Injury

Potatoes can be damaged by herbicides applied during the season, by herbicide residues that carry over from previous crops, or by drift from a neighboring area. Herbicide injury is characterized by yellowing or distorted growth. Laboratory analysis of soil, water, or plant residues may be required to verify an herbicide-related injury. However, presence of an herbicide residue is not necessarily proof that it caused the injury. Analysis is

AL MOSLEY

The youngest leaves of plants injured by glyphosate appear wilted and turn yellow.

Tubers of plants injured by glyphosate may develop folds or clefts and a rough surface.

Aerial tubers may form at the base of stems or in leaf axils when disease or physiological stress interrupts the flow of carbohydrates from leaves to tubers. Diseases that may cause aerial tuber formation include blackleg, Rhizoctonia, aster yellows, witches' broom, and curly top virus.

Injury by plant growth regulator herbicides, such as 2,4-D, dicamba, or picloram, appears as roughened, crinkled leaves. With severe injury, "fiddleneck" develops—young stems are curled and leaves are severely distorted or absent.

Ozone causes spots or patches of dead tissue on the leaves of sensitive potato varieties. This symptom is often called speckle leaf.

easier if you can tell the laboratory what materials have been applied; keep records for all fields of all materials and rates used during rotations as well as on the potato crop.

The classes of herbicides that may damage potatoes are listed below, along with descriptions of the symptoms they cause. Use these descriptions to help identify herbicide injuries, but also check with local extension agents, farm advisors, or other experts.

Acetanilides: metolachlor

Metolachlor is soil-applied to potatoes before emergence, and may cause injury under some circumstances. Injured leaves appear bronze; leaf margins curl up or down and turn brown. Leaf surfaces appear crinkled or roughened. Occasionally leaves are small and odd shaped with roughened surfaces.

Bromoxynil

Injury results when bromoxynil spray drifts onto potato foliage during the treatment of adjacent grain fields. Localized areas on sprayed leaves turn bronze, then yellow. Leaf margins may curl up and turn brown, and interveinal tissue may turn brown.

Dinitroanilines: pendimithalin, trifluralin

Dinitroanilines are applied to the soil before potatoes emerge. Injury may occur if sprouts grow slowly through treated soil that is cold and wet. Symptoms are more frequently caused by trifluralin than by pendimithalin. Developing shoots become thick and brittle, root formation may be slowed or inhibited, and leaves may become crinkled and cupped.

Glyphosate

Drift of glyphosate spray onto potato foliage causes injury. Youngest tissue is affected first. Petioles bend downward and leaf margins curl slightly, giving growing points a wilted appearance. The youngest leaflets turn yellow. In severe cases older foliage is affected and plants may die. Tuber malformations sometimes occur. A fold or crease may develop, usually at the bud end of tubers. Roughened surface indentations resembling grub damage may also develop.

Growth Regulators: 2,4-D, 3,6-DPA, dicamba, picloram

Dicamba and 2,4-D injuries are caused by spray drift; 3,6-DPA and picloram injuries result from residue carry-over in the soil. Dicamba residues are not a problem when the herbicide is applied at normal rates at least 6 months before potatoes are planted. Symptoms appear on actively growing tissue. Margins of the youngest leaves curl or the leaves fold up. Leaflets appear roughened or crinkled, and petioles bend downward. With severe injury, young stems are curled and leaves are thin and severely distorted or absent; these symptoms are often called "fiddleneck." Dicamba and picloram may cause tuber malformations. If dicamba spray injury occurs during tuber growth, tubers may be rounder than normal, a cleft may develop at the bud end, and the surface of the bud end may develop a roughness similar to elephant hide.

Thiocarbamates: EPTC

Potato foliage may be injured if EPTC is applied at too high a rate or if applications overlap during preplant soil incorporation. Shoots appear twisted, and leaves become cupped, curled, or bent over.

Triazines: atrazine, hexazinone, metribuzin, simazine

Atrazine, hexazinone, and simazine usually cause damage when residue remains from treatment of a previous crop. Metribuzin may cause injury when high rates are used, when it is applied to sensitive cultivars, or when poor growing conditions (cloudy, cool, wet weather) occur before or after it is applied. Metribuzin injury is more likely following foliar applications. Triazine injury symptoms from soil uptake appear in larger, older leaves; symptoms from a foliar application are confined to the leaves sprayed. Leaves turn yellow either along the veins or between the veins, depending on cultivar. If injury is minor, the yellow tissue will turn green; if severe, yellow tissue eventually dies, and leaf margins take on a burned appearance.

Uracils: terbacil

Terbacil injury is usually caused by residue carryover from previous crops. Symptoms are similar to those caused by triazines.

Ureas: chlorsulfuron, diuron, linuron

Symptoms of urea injury are similar to those of triazine injury, and are usually caused by residue carryover from previous crops. Chlorsulfuron or diuron injury may also result from spray drift. Chlorsulfuron is generally not recommended for use in rotations where potatoes will be planted, because the waiting period for potatoes is at least 3 or 4 years.

Nutrient Deficiencies

Nutrient deficiencies may be caused by low levels of nutrients in the soil and factors that reduce the ability of the potato plant to absorb nutrients. Excessive rain or irrigation can leach mobile nutrients such as nitrogen from the root zone, soil pH or nutrient imbalance can decrease the availability of certain nutrients, and disease may reduce the ability of roots to absorb nutrients. Deficiency symptoms include abnormal or reduced growth, discoloration of foliage, reduced yields, and improper tuber development. Plants deficient in nutrients are more susceptible to diseases and disorders such as Verticillium wilt, early blight, bruising, and internal browning. Use soil and tissue analysis to assess the nutrient status of your soil and crop, and follow fertilizer recommendations for your area and the cultivars you grow to avoid nutrient deficiencies. Identification of nutrient deficiencies may be difficult. Use the results of soil and tissue analyses and seek the advice of experts before taking corrective action.

Magnesium

Symptoms of magnesium deficiency develop on older, lower leaves. Interveinal tissue of leaflet margins and tips turns pale and the discoloration progresses toward the center of the leaflet, where symptom expression becomes most severe. Discolored tissue turns brown and leaves become thick and brittle, and roll upward. Magnesium deficiency usually occurs on acidic, sandy soils, from which the nutrient is easily leached; however, magnesium deficiency may also occur on finer-textured soils. Symptoms often appear after heavy rains, which leach magnesium from the root zone. High levels of potassium in the soil aggravate magnesium deficiency. Foliar sprays of magnesium sulfate or chelated micronutrients may be used to correct deficiencies that appear during the season.

Manganese

Manganese deficiency symptoms develop on the upper part of the potato plant. Interveinal leaf tissue first loses its luster, then turns pale. Discoloration progresses until the tissue becomes yellow or white. If manganese deficiency is severe, brown spots may develop along the veins of younger leaves. Manganese deficiency occurs most often in muck soils or calcareous, alkaline soils. Foliar

applications of manganese sulfate or chelated micronutrients may be used to correct deficiencies during the season.

Nitrogen

Potato plants deficient in nitrogen grow more slowly, form small, pale leaves, and turn yellow. Leaf veins stay green longer as leaves turn yellow. Older leaves are affected more severely. Nitrogen deficiency reduces yields and makes plants more susceptible to Verticillium wilt, early blight, and other diseases. Nitrogen deficiencies can be corrected by midseason fertilizer applications, but some loss of yield and tuber quality may still occur.

Phosphorus

Early season phosphorus deficiency causes leaves to curl and turn a purplish color. Phosphorus deficiency late in the season may cause premature maturation, increased disease susceptibility, and reduced yields. Phosphorus deficiencies are difficult to correct during the season. Use preplant soil analysis and results of petiole analysis from the previous potato season as guides to determine adequate phosphorus application before and during planting.

Potassium

Leaves of plants deficient in potassium develop dark green to bluish color early in the season. Older leaves turn brown at their outer edges, become bronzed in appearance, and die early. Margins of leaves in the upper plant curl upward, and the leaflets become bronzed. Potassium-deficient plants are more susceptible to early blight and Verticillium wilt; their tubers are more susceptible to blackspot bruise and early blight infections.

Sulfur

Yellowing of potato foliage throughout the plant may indicate a sulfur deficiency . The intensity of the yellowing varies from slight to severe. Leaflets may develop some upward curling. Sulfur deficiency may occur where soil is low in sulfur and levels are low in irrigation water.

Zinc

Potato plants suffering from zinc deficiency are stunted and have younger leaves that turn yellow and roll upward, symptoms that are similar to current-season leafroll. Bronze areas that later turn brown and die may appear on leaves. This symptom develops first on leaves in the middle of the plant and spreads throughout the

Tubers of plants injured by dicamba or picloram may be rounder than normal with clefts at the bud end and rough surfaces.

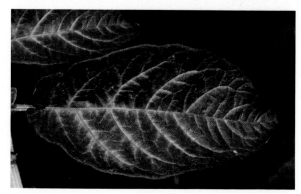

Triazine herbicides, such as atrazine, hexazinone, simazine, or metribuzin, may cause yellowing of tissue along the leaf veins of some cultivars. In other cultivars leaf tissue turns yellow between veins.

plant later. Brownish spots may appear on stems and petioles. Zinc deficiency is most commonly seen on newly developed land and on alkaline soils. Excessive liming and high phosphorus levels accentuate zinc deficiency. Foliar applications of zinc chloride, zinc sulfate, or chelated micronutrients can be used to correct zinc deficiency.

Leaves of plants with manganese deficiency turn pale. Brown spots develop along the veins of younger leaves when the deficiency is severe.

Magnesium deficiency causes yellowing of interveinal leaf tissue. Brown spots develop and leaves thicken and roll upward at their edges.

Nitrogen deficiency causes leaves to turn a pale yellow green. Older leaves are usually affected more severely.

ALBERT ULRICH

Phosphorus deficiency causes leaves to curl and turn a purplish color.

ALBERT ULRICH

Leaves of plants deficient in potassium develop a dark green to bluish color. Upper leaves curl; lower leaves turn yellow at the edges and brown patches develop.

ALBERT ULRICH

Young leaves of plants suffering from zinc deficiency turn pale and roll upward. Lower leaves develop bronze areas, turn brown, and die.

Nematodes

Nematodes are tiny roundworms that live in soil, water, and plant tissues. Many damage plants, but others feed on fungi, bacteria, or other nematodes. In the western states root-knot nematodes cause the most damage to potatoes. Root-lesion nematodes, which are found in many potato fields, may increase the severity of early dying and may reduce yields if present in high numbers. Stubby-root nematodes are widespread in the western states; they transmit tobacco rattle virus, which causes corky ringspot. The potato-rot nematode is known in the West but occurs rarely. Potato cyst nematode does not occur in the western states, but growers should be aware of its symptoms because it might be introduced on contaminated seed tubers.

Cropping history, soil sampling results, and past experience in local soils must all be considered when making nematode management decisions. Where soil populations of a nematode are high enough to cause unacceptable damage, plant nonhost crops or use fumigation or other soil treatments to reduce the nematode to levels that will not cause economic damage. Sanitation and use of certified seed help prevent spread to uninfested fields. Crop rotation with good weed control helps prevent established populations from increasing.

□ *M. chitwoodi*

▨ *M. chitwoodi + M. hapla*

▩ *M. hapla + M. incognita*

■ *M. chitwoodi + M. hapla + M. incognita*

▢ *M. incognita*

▨ *Meloidogyne* species not identified

Figure 36. Distribution of major root-knot nematode species in the potato growing areas of the western United States (as of February, 1986).

Root-Knot Nematode
Meloidogyne spp.

Root-knot nematodes occur throughout the potato-growing areas of the western states and are presently serious pests in California, Oregon, Washington, and Idaho. Most crop plants are susceptible to one or more species of root-knot nematode. Several species of *Meloidogyne* are known to attack potato, three of which are of major importance. All have similar life cycles, but the species differ in environmental requirements, host range, and some of the symptoms they cause. The distribution of each nematode species (Figure 36) is determined primarily by its temperature requirements (Table 13). The Columbia root-knot nematode, *M. chitwoodi*, and the northern root-knot nematode, M. *hapla*, occur in the cooler growing areas; M. *chitwoodi* is more active than M. *hapla* at lower temperatures. The southern root-knot nematode, M. *incognita*, is the predominant species in the San Joaquin Valley and southern California, Arizona, and New Mexico. *Meloidogyne hapla* is known to occur in these warmer areas, and occasionally fields will be infested with Javanese root-knot nematode, M. *javanica*, or peanut root-knot nematode, M. *arenaria*. Two or more species are often found together, but one usually predominates.

Table 13. Temperature Requirements of the Three Major Root-Knot Nematode Species That Infect Potatoes.

| | SOIL TEMPERATURE AT ROOT OR TUBER DEPTH FOR SPECIES | | |
	M. *chitwoodi*	M. *hapla*	M. *incognita*
Minimum for reproduction	<45°F (6°C)	<54°F (12°C)	<50°F (10°C)
Minimum for infection	<45°F (6°C)	<54°F (12°C)	<65°F (18°C)
Optimum for activity	68°–77°F (20°–25°C)	77°–86°F (25°–30°C)	76°–90°F (25°–32°C)

Losses from root-knot nematode damage to tubers may be reduced by early harvest. In winter and spring production areas, damage may be avoided by harvesting before nematode populations build up to damaging levels. Losses due to Columbia root-knot nematode may also be reduced by avoiding storage of tubers. Root-knot nematode populations may be controlled with rotations to nonhost crops, dry fallowing, tillage to increase exposure to extreme temperatures, and nematicides.

Description and Biology

Root-knot nematodes are parasites that must feed inside roots or tubers to reproduce. All common species attack a wide variety of crops and weeds (Table 14); feeding by most species causes host plants to produce swellings, called galls or knots, around the feeding sites (Figure 37). These abnormalities prevent normal uptake of water and nutrients.

Mature females, found only in roots or tubers, are pear shaped and up to about 1/16 inch (1.5 mm) long. They can sometimes be seen as tiny white "pearls" when infected roots are cut open or when slices of infected tubers are held up to the light. Males are worm shaped and under normal conditions are not abundant; they are not necessary for reproduction. Females lay eggs in jelly-like masses on or just below the surface of infected roots and inside infected tubers; tuber cells surrounding egg masses turn brown. Eggs inside tubers can survive winter conditions; a small proportion of M. *chitwoodi* eggs

can survive for more than 2 years inside tubers held at 34°F (1°C).

The worm-shaped juveniles that hatch from eggs can move as far as 2 or 3 feet through moist soil, but usually travel only short distances to find a host plant. They penetrate roots just behind root tips, where the root surface has not been strengthened with age (suberized). The site of tuber infection is not known for certain; juveniles probably enter through lenticels or eyes, but they also may penetrate directly through the skin. Mature tubers with well-developed skin appear to be infected less than immature tubers. Inside the root or tuber, the nematode's salivary secretions contain enzymes and plant

Table 14. Some Hosts of the Major Species of Root-Knot Nematodes That Infest Potatoes.[a]

| | ROOT-KNOT SPECIES | | |
	Columbia M. *chitwoodi*	Northern M. *hapla*	Southern M. *incognita*
Crop			
Alfalfa	+[b]	+	+
Beans	+	+	+
Carrots	+	+	+
Cole crops	+	+	+
Corn	+	−	+
Cotton	−	−	+
Eggplant	−	+	+
Grains	+	−	+
Grapes	−	+	+
Hops	−	−	0
Lettuce	0	+	+
Mint	−	+	−
Melons	−	+	+
Peas	+	+	+
Peppers	−	+	+
Strawberry	−	+	0
Stone fruit	−	−	+
Sudangrass	−	−	−
Sugar beet	+	+	+
Tomato	+	+	+
Weed			
Barnyardgrass	−	−	−
Bindweed	0	+	+
Canada thistle	+	+	+
Foxtails	−[c]	−	+
Kochia	0	+	0
Lambsquarters	−	+	+
Mallow	0	+	+
Mustards	−	+	+
Nightshades	−	+	+
Nutsedge	0	0	+
Pigweeds	−	−	+
Russian thistle	−	0	0
Sowthistle	+	+	+

a. (+ = good host, − = poor or nonhost, 0 = unknown)
b. Alfalfa is a host for race 2 of M. *chitwoodi*.
c. Green foxtail is a moderate host, yellow foxtail and meadow foxtail are poor or nonhosts.

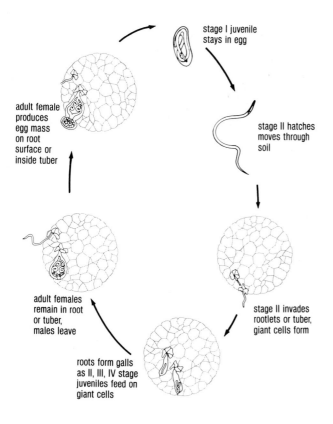

stage I juvenile
stays in egg

adult female
produces
egg mass
on root
surface or
inside tuber

stage II hatches
moves through
soil

adult females
remain in root
or tuber,
males leave

stage II invades
rootlets or tuber,
giant cells form

roots form galls
as II, III, IV stage
juveniles feed on
giant cells

Figure 37. The root-knot nematode spends much of its life cycle in roots or tubers. Second stage juveniles invade new sites, usually near root tips or tuber lenticels or eyes, causing some cells to grow into giant cells where the nematodes feed. As feeding continues, tubers produce lumps; roots produce small galls in response to some species.

hormones that stimulate the formation of "giant cells," greatly enlarged cells that supply the nematode with food. Other cells may multiply to form swellings or galls, but potato roots infected by M. chitwoodi do not form galls. As they mature, females swell, become pear shaped, and lose the ability to move.

When susceptible host plants are available, a root-knot population increases at a rate that depends largely on soil temperature and moisture and the number of nematodes present in the spring. Populations usually decline by over 50% during the winter; spring populations consist mostly of eggs and juveniles. Meloidogyne chitwoodi is active at lower temperatures than the other two species; therefore, it builds up to damaging levels earlier in the season because it has more generations and is more active on young plants in cooler soils. In areas where M. incognita is the important species, most potatoes are planted in winter and harvested in late spring.

Because this species does not infect until soil temperatures reach 65°F (18°C), tubers are usually harvested before the nematode invades and causes damage.

Root-knot nematodes usually are concentrated in the upper 2 feet of soil where feeder roots are abundant, but they can be found as deep in soil as roots penetrate; M. chitwoodi has been found as deep as 6 feet.

Meloidogyne chitwoodi can reproduce in stored tubers at temperatures as low as 46°F (8°C), thereby increasing damage during storage. Increase of tuber damage by other species in storage has not been reported.

Symptoms and Damage

Root-knot symptoms occur on tubers and roots; symptoms on tubers are more common and easier to see. Meloidogyne chitwoodi and M. incognita cause characteristic bumps or warts on the surfaces of infected tubers. Meloidogyne hapla causes swellings that are less distinct; it may be hard to distinguish these tubers from healthy ones just by observing surface symptoms. Root-knot nematodes are usually located in the tuber cortex, between the skin and vascular ring. Sometimes they penetrate deep into the tuber, in which case no surface symptoms are visible.

Brown spots develop in tubers around the egg masses of all root-knot species, which occur mostly in the outer 1/4 inch (6 mm) and are best seen by carefully peeling off thin layers of tuber. The spots darken when tuber slices are cooked to make chips; therefore, tuber lots that are significantly infected with any nematode species are unacceptable for processing. Warty tubers cannot be used for fresh market.

Root-knot galls on feeder roots are usually small and difficult to see. Meloidogyne chitwoodi produces egg masses that appear as tiny round bumps on feeder roots of heavily infected plants. Meloidogyne hapla causes small, distinct galls and proliferation of lateral roots around these galls. Meloidogyne incognita can cause more pronounced root galls, but it usually is not active in cool weather, when most potatoes are grown in the areas where it occurs.

Root-knot infections rarely cause aboveground symptoms in potatoes. However, plants with severe root-knot infections may be stunted, turn yellow, wilt, or die if stressed for water.

Management Guidelines

Management decisions for root-knot nematodes must be made well in advance of planting. Fumigation, when effective, works best in the fall, so it is important to keep records from the previous potato crop and to monitor in the season before planting. Check for nematode symptoms before harvest and keep records of where they are found; field location of tuber symptoms in the last potato crop give a good indication of where to sample for root-knot nematodes. Egg masses on the surface of root samples taken at midseason are easily seen by staining with Phloxine B. Some weed hosts can be good indicators of root-knot. *Meloidogyne chitwoodi* produces obvious galling on bull thistle, *Cirsium vulgare*, and spiny sowthistle, *Sonchus asper*. *Meloidogyne hapla* produces distinct galls on dandelion, *Taraxacum officinale*; *M. incognita* on nightshades and ground cherries. Symptoms on weeds or previous crops are a good sign that potatoes planted in the same soil are likely to be infected; however, the absence of root-knot symptoms does not necessarily mean that root-knot nematodes are not present or that a potato crop will not be damaged. Some crops and weeds either are not susceptible to infection or do not develop distinctive symptoms; or there may be root-knot nematodes in soil when no suitable host is present.

Tomatoes are sometimes grown in soil samples to test for the presence of root-knot nematodes. If you use galling on tomatoes to indicate the presence of *Meloidogyne* spp., avoid Red Cherry or Rutgers cultivars, which produce almost no galling in response to *M. chitwoodi*. Roza, Yellow Pear, Columbia, and Saladmaster do produce gall-like symptoms in response to this root-knot species. Other species will induce gall formation on all susceptible tomato cultivars.

Differential Host Test. To make decisions about crop rotations and controls, it is important to know which root-knot species are present. Root-knot nematodes isolated from soil samples can be identified, but this requires experience and a good microscope. The differential host test, which uses indicator plants susceptible to only one root-knot species, is a simpler way to identify nematodes. The test uses the Nugaines cultivar of wheat—a good host for *M. chitwoodi* but a nonhost for *M. hapla*—and California Wonder cultivar of pepper—a good host for *M. hapla* but a nonhost for *M. chitwoodi*. In areas where *M. hapla* and *M. incognita* occur, but not *M. chitwoodi*, substitute Charleston Grey watermelon for Nugaines wheat. Follow the steps below to carry out the test.

1. Mix a soil sample from the field to be tested (see *Soil Sampling*, below), then divide it into four 4- to 6-inch pots. Plant one to three wheat seeds in two of the pots and one to three pepper seeds in the other two pots. Grow the plants at 70° to 80°F (21° to 27°C) for 55 days.
2. Remove the plants from the pots and immerse each root system in fresh water to gently wash out the soil. Remove and discard the tops. Rinse the container after each root system. Blot each root system with a paper towel to remove excess water. Place each root system in a separate, watertight plastic bag.
3. Prepare a Phloxine B stock solution by adding 1/2 ounce (14 g) of Phloxine B to 2 1/2 gallons (10 liters) of tap water. Label this stock solution and store at room temperature. (Phloxine B can be obtained from sources listed in the References.)
4. Mix one part of the stock solution with nine parts of tap water. Cover the roots in the watertight plastic bag with this solution, wait 15 minutes, then pour off the liquid. Roots will stain a light pinkish orange; egg masses of root-knot nematodes will stain a dark red.
5. Check for egg masses on each stained root system with a magnifying lens or a dissecting microscope, and rate each plant as either a host, if egg masses are present, or a nonhost, if egg masses are absent.

If wheat plants are hosts, then *M. chitwoodi* is present; if pepper plants are hosts, *M. hapla* is present. Both nematodes may be present in the same soil sample. Watermelon indicates the presence of *M. incognita*, which causes the formation of large, distinctive galls. *Meloidogyne chitwoodi* may also cause gall formation on this host, but the galls are tiny in comparison to those caused by *M. incognita*.

The Phloxine B stain can also be used to test samples of potato roots taken from the field. The stain will indicate the presence of root-knot nematodes but will not differentiate between species.

Soil Sampling. A more precise way to monitor for nematodes before a new crop is planted is to take a series of soil samples and submit them to a qualified laboratory for extraction and, if necessary, identification of the nematode species present. Samples should be taken in the fall after harvest or preferably when the previous crop is still in the field; nematodes normally are concentrated in the root zone of plants, and it is easier to find them when the plants are present. It is important to sample while the soil is in good working condition and is neither saturated nor very dry. Contact the laboratory in advance and schedule sampling so that the lab can process the samples as soon as possible after they are received. Farm advisors, extension agents, or other authorities can help you find a laboratory equipped for extracting and identifying nematodes from soil samples. Take the following steps, but also contact the lab for any special instructions they may have. For more information on nematode sampling, consult *Sampling for Nematodes in Soil* listed in the References.

1. Draw a map of the field, showing areas that differ from the rest in soil texture, cropping history, or crop injury (Figure 38).
2. Divide the area into blocks of 5 acres or less, using a grid pattern. If conditions are uniform throughout the field, apply the grid pattern to the whole field.

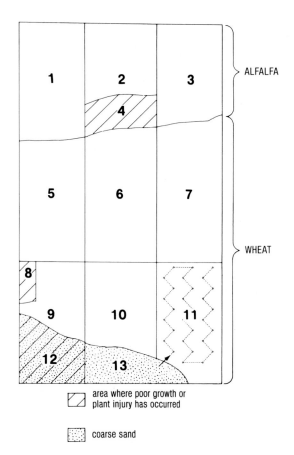

area where poor growth or plant injury has occurred

coarse sand

Figure 38. Before taking soil samples or checking roots for galls, divide the field into areas that reflect any differences in cropping history or soil type. For soil sampling, subdivide these areas into blocks of about 5 acres. If there are areas where there has been poor growth, treat them as separate blocks, even if they are smaller than other blocks. Assign each block a number and collect soil from a series of points in each one, following a systematic pattern as shown in block 11.

In general, the smaller the blocks are, the more precise the result of sampling will be.

3. Collect 20 or more subsamples of soil and roots from each block with a shovel or with a soil auger or Oakfield tube that takes a core 1 inch (2.5 cm) or more in diameter. Take the soil from the root zone of the crop to a depth of 18 inches (45 cm) or more; take deeper samples from bare ground. Include as many feeder roots as possible. Make sure the soil is moist, but avoid places where soil is wet or compacted. If the surface soil is dry, discard it and include only moist soil.

4. Mix the subsamples from a single block thoroughly in a bucket or bag, then transfer about 1 quart (1 liter) into a plastic bag or other moisture-proof container.

5. Label the sample *in pencil* with the block number. Put the label on the *outside* of the container, since moisture may ruin labels placed inside. Label all samples with your name and address, the date, location of the field, crop history, the crop present when you took the samples, the crop you intend to plant, and any notes you have of previous crop injury.

6. Keep the samples cool; do not leave them in the sun and do not freeze them. The best storage temperature is 50° to 60°F (10° to 15°C). Seal the containers so they will not dry out. Samples may be kept in good condition by keeping them in an ice chest until they arrive at the laboratory.

7. Send or deliver the samples to the lab immediately in a cardboard box insulated with newspaper or in an ice chest packed to keep samples from breaking in transit.

It usually takes about 2 weeks for a lab to perform a routine analysis for nematodes. Methods commonly used to extract root-knot nematodes from soil recover only juveniles, and the results are usually reported as the number of nematodes found in a standard volume or weight of soil, such as 250 cc or 1 kg. The centrifugal-flotation method and the Baermann funnel method are most commonly used for nematode extraction. Each lab report *should specify the extraction method used*, and it *should also indicate the estimated proportion of the nematodes present that are actually extracted*. This figure, called the *recovery* or *efficiency rate*, is usually from 10% to 30% for root-knot juveniles.

Laboratories should identify to genus and list other plant parasitic nematodes such as root-lesion nematode (*Pratylenchus* spp.) and stubby-root nematode (*Paratrichodorus* or *Trichodorus* spp.), and will also identify, if requested, the species of root-knot nematode that are found. Species identification requires rearing the juveniles to the adult stage on susceptible hosts, which increases the cost of lab analysis but can be valuable in choosing rotation crops and planning control procedures. Species identification is generally necessary only where M. *hapla* and M. *chitwoodi* may both occur.

Besides root-knot nematodes, many other kinds of plant parasitic nematodes are often found in soil samples. Commonly found in potato soils are root-lesion nematodes, stubby-root nematodes, stunt nematodes (*Tylenchorhynchus* spp.), pin nematodes (*Paratylenchus* spp.), and spiral nematodes (*Helicotylenchus* spp). Root-lesion nematodes and stubby-root nematodes may be of economic importance; none of the others are important on potatoes.

Control

There are no precise guidelines for deciding what number of root-knot nematodes found in soil samples will cause unacceptable damage in potatoes. If any M. *chitwoodi* are found in samples, damage can be expected. Generally, if root-knot nematodes are found in fields to be planted to potatoes in the spring and the area has a history of nematode damage to potatoes, soil treatments should be used to reduce the nematode populations. In areas where potatoes are planted in winter and harvested in spring, tubers usually escape significant damage even if root-knot nematodes are present. Check with specialists in your area for treatment recommendations based on the results of soil sampling.

Control measures for root-knot nematodes include the use of sanitation, crop rotation, dry fallow, and nematicide treatments. The choice depends on the cost of these measures, the timing of control decisions relative to season and weather, the value of the crop, and the expected impact of the nematode population under local conditions. Research is underway to develop potato cultivars resistant to root-knot nematodes; however, none are available at the present time.

Sanitation. Root-knot nematodes do not move far by themselves, so undisturbed infestations spread slowly. However, they can be spread in infected tubers, in soil carried on farm equipment, and in reused irrigation water. A small infestation in one part of a field can easily spread throughout the field when soil is moved in grading. Some species have been widely distributed by the shipment of infested soil and infected tubers.

Follow the sanitation measures listed in the chapter, *Managing Pests in Potatoes*, to avoid spreading infestations of root-knot nematodes. The most important points are to use clean seed tubers, avoid moving infested soil or fresh manure or organic matter, and avoid using contaminated water. Use of settling ponds will reduce the spread of nematodes in surface irrigation water. Leave tubers that remain after harvest exposed to the environment; this helps reduce survival of both nematodes and potato volunteers.

Crop Rotation. Root-knot gets worse and control more difficult where susceptible crops are grown without rotation. Crop rotation will not eliminate infestations because small populations of most root-knot species survive in the soil as eggs for up to 2 years in the absence of host crops, and most species also reproduce on a wide range of weeds (Table 14). Also, almost all rotation crops are hosts for M. *incognita* and for either M. *hapla* or M. *chitwoodi*. However, rotations can be used to reduce the soil population of the dominant nematode species in a given field and make other control measures more effective. Local extension agents, farm advisors, or other experts can help choose suitable rotation crops. Managing weeds during and between rotations is essential for controlling root-knot nematodes. In some situations nematicides may be used on rotation crops to reduce nematode populations.

In areas where both M. *hapla* and M. *chitwoodi* nematodes are present, identification of the species found in soil samples is important in making rotation decisions. Of the crops commonly rotated with potatoes, cereals are nonhosts for M. *hapla* but good hosts for M. *chitwoodi*, and would be the best rotation choice if M. *hapla* is the dominant species. Alfalfa is a good host for M. *hapla* and for race 2 of M. *chitwoodi*, which is widespread in the western states. Research is underway to develop crop cultivars that are resistant to both nematode species.

In the San Joaquin Valley, southern California, Arizona, and New Mexico, where M. *incognita* occurs, soil fumigants or nematicides are sometimes applied to control nematodes in rotation crops such as tomatoes, lettuce, carrots, sugar beets, or cotton. However, nematode populations may build up to damaging levels within one season after a nematicide treatment if susceptible crops are grown. Winter cereals are commonly used for rotation in these areas. Use of crops such as certain processing tomato cultivars that are resistant to M. *incognita*, M. *javanica*, and M. *arenaria* will reduce population levels and will reduce the potential for damage to potatoes. Avoid planting winter potatoes following a crop that had severe root-knot nematode damage.

Early Harvest. The sooner tubers are harvested the less they are damaged by root-knot nematodes. Where winter potatoes are harvested in April or May, damage is usually avoided because populations of M. *incognita* are still low; if tubers are left until July, nematode damage can be severe. Likewise, early potato cultivars are seldom damaged by M. *chitwoodi* or M. *hapla* if tubers are harvested before August. In all growing areas, if root-knot nematodes are present, harvest tubers as soon as possible after allowing proper maturation; nematode

Root-knot nematodes cause various types of swellings, bumps, and warts on tuber surfaces, depending on the root-knot species (*tuber at left is uninfected; tuber at right is infected with Columbia root-knot nematode*). All species cause brown spots in the cortex of the tuber (*below*).

Root-knot nematode egg masses on root surfaces are stained dark pink by Phloxine B.

damage increases the longer tubers are left in the ground, even if nematicide treatments were made before the crop was planted.

Fallow Cultivation. Repeated cultivation of fallow land reduces nematode populations in the upper 12 to 18 inches (30 to 45 cm) of soil by exposing them to heat and to the drying action of the air. In areas where winter potatoes are grown, dry fallow during the summer helps keep nematode populations from increasing. Follow recommended weed control practices to reduce the number of weed hosts.

Nematicides. Chemicals can be used to reduce populations of root-knot nematodes to below damaging levels, although they will not eradicate them. Nematode popu-

lations are likely to build back up to economically damaging levels within 1 year after treatment if host crops are planted. Nematicides are generally less effective in controlling M. *chitwoodi* because it can build up more quickly than other root-knot species.

All soil fumigants now available are toxic to crops, so a waiting period is required between application and planting. This period is at least 10 days in most cases and may be 3 weeks or more, depending on the material and rate used and soil conditions. Always observe the waiting period indicated on the label. Fumigating in the fall is usually best because the waiting period before planting is usually not a problem and because spring fumigation may be less effective due to cool, wet soil.

Proper soil moisture and temperature are needed for controlling nematodes with fumigants. Fumigants do not disperse properly in cold, wet soil, so treatments are less effective and the waiting period before planting is longer. Soil that is too dry may allow the fumigant to escape too fast, especially in warm weather. Dry soil does not allow adequate penetration of fumigants applied through sprinklers. Correct moisture levels depend on the fumigant used; follow label recommendations.

Check the soil temperature at the depth of application with a dial thermometer before fumigating. The best temperature for fumigating is 50° to 70°F (10° to 21°C). Do not fumigate if the temperature is below 40°F (4.4°C) or above 80°F (27°C). The cooler the soil, the more important that it not be too wet.

Good soil preparation and proper application techniques are also important. Application rates depend on soil type, and procedures are affected by wind; check label directions. Before applying a soil-injected fumigant, prepare the soil to seedbed condition by deep plowing and discing or rototilling so that the soil is worked up to the desired fumigation depth. Be sure clods are broken up and crop residues are decomposed; fumigants may not penetrate them and nematodes in intact plant parts are more resistant to nematicides.

Take a soil sample after fumigation and before planting to check the effectiveness of a fumigation. Depending on the materials and methods used, fumigation may also destroy weeds and kill soilborne plant pathogens such as *Verticillium*.

Certain nonfumigant nematicides will control M. *hapla* but not M. *chitwoodi*. They are frequently used in combination with the soil fumigants to control M. *chitwoodi*, especially when populations are high or the nematode is present deep in the soil.

Check the References in the back of this manual for information about specific chemicals and application procedures that can be used in your area. Also, consult your local extension agent, farm advisor, or other experts for specific nematicide recommendations. *Always follow label directions carefully when applying any of these chemicals.*

Root-Lesion Nematode
Pratylenchus spp.

Several species of root-lesion nematode are found throughout the western potato-growing areas, but they usually do not cause economic damage. All feed on roots and most do not attack tubers. Two species, *Pratylenchus penetrans* and *P. neglectus*, can increase the susceptibility of potato plants to Verticillium wilt.

Root-lesion nematodes lay their eggs inside the roots of infected plants or in the adjacent soil. The larvae that emerge from these eggs infect roots just behind the growing tips, causing reddish brown lesions around the root cortex. These lesions grow together and turn black; under field conditions they may be difficult to see. The lesions are often invaded by soil microorganisms, and yields may be reduced if root damage is extensive. Damage thresholds have been established in other parts of the United States by relating yields to soil populations of *Pratylenchus*. Such thresholds are not presently available for western growing areas.

Some species of *Pratylenchus* cause serious damage on potato tubers, however they are not known to occur in the western United States. Tubers can be attacked by *P. neglectus*, but economic damage to tubers is not generally a problem with this species in the western states.

Crop rotation is not an effective means of controlling root-lesion nematodes, because they have a wide host range. They are particularly favored by grain crops. Follow the sanitation recommendations for root-knot nematodes to prevent spread of root-lesion nematodes. Fumigation for root-knot nematodes will control root-lesion nematodes as well, but fumigation is usually not recommended to control *Pratylenchus* alone. If root-lesion nematodes are causing a problem, nonfumigant nematicides will usually control them. Consult your extension agent, farm advisor, or other experts for specific recommendations.

Stubby-Root Nematode
Paratrichodorus (*Trichodorus*) spp.

Though widespread in the western states, stubby-root nematodes are of economic concern only in certain areas of California, Colorado, Idaho, Oregon, and Washington, where they are vectors for tobacco rattle virus (TRV), the virus that causes corky ringspot of potato tubers. Stubby-root nematodes are confined to coarse, sandy soils, in which they can live as deep as 40 inches (1 m) or more. They attack the roots of a wide range of plants, feeding on root surfaces and causing the formation of numerous stubby roots. Root damage does not affect potatoes significantly; the nematodes become a problem when they transmit tobacco rattle virus.

A large number of crops and weeds are hosts of tobacco rattle virus. Stubby-root nematodes acquire the virus when feeding on infected roots, and they transmit the virus when they move to a new host. Corky ringspot develops after virus-infected nematodes feed on developing potato tubers.

Avoid moving soil from fields where corky ringspot is found; if possible, do not plant potatoes in these fields. In some states, fields where corky ringspot has been found must be permanently prohibited from growing certified seed potatoes.

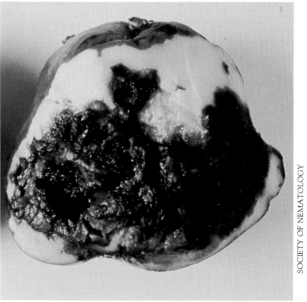

SOCIETY OF NEMATOLOGY

Potato-rot nematode causes dark brown, honeycombed, granular rot. Rotted tissue becomes wet and soft if invaded by bacteria.

BILL BRODIE

Female potato cyst nematodes form egg-containing cysts, which are 1/50 to 1/30 inch (0.5 to 0.8 mm) in diameter and contain up to 500 eggs, on the surfaces of host plant roots. The potato cyst nematode does not occur in the western states.

If control is necessary, nonfumigant nematicides are usually more effective than fumigants because stubby-root nematodes deep in the soil may escape the fumigant and move upward after the fumigant has dissipated. For best results with nonfumigants, apply them at planting, when most stubby-root nematodes are in the upper soil layers, and be sure the material is not washed out of the upper layer of soil by irrigation or rain. When applied at planting, nonfumigants will remain effective long enough to prevent damage during tuber initiation and growth by nematodes that may move up from deeper soil layers.

If you want to determine if stubby-root nematodes are present in fall soil samples, be sure to sample to a depth of at least 24 inches (61 cm), because these nematodes move deep into the soil at that time of year.

Potato-Rot Nematode
Ditylenchus destructor

In the West the potato-rot nematode is known to cause damage to potatoes only in a few isolated areas of Idaho. The nematode overwinters as eggs and survives on a large number of plants including grains, tomatoes, carrots, bulbous lilies, and several weed species. *Ditylenchus* is more frequently a problem in the production of bulbous iris or dahlia bulbs than in potatoes.

Juveniles hatch from eggs and attack stolons and tubers, entering tubers through eyes or lenticels. Symptoms appear only on tubers, first as pinhole-sized lesions that may be confused with wireworm damage. The nematodes can be seen just below the skin as pearly white spots. Lesions usually develop in storage; the nematodes are active at temperatures as low as 40°F (4.4°C). Cavities develop around the nematodes, and the tuber surface becomes dry and cracked. The rot usually takes on a chalky, honeycombed appearance. If the chalkiness is not apparent, symptoms may be confused with scab or tuber decay caused by ring rot or late blight, but nematodes are easily seen in rotted tissue examined with a dissecting microscope.

Ditylenchus may be controlled by soil fumigation. Avoid planting potatoes where the nematode occurs. Avoid moving infested soil or infected tubers. Do not store tubers if symptoms are seen during harvest.

Potato Cyst Nematode
Globodera rostochiensis

The cyst nematode, also called the golden nematode, occurs in some potato-growing areas of Canada and the eastern United States. Strict quarantines and seed certification requirements are used to prevent it from spreading to other potato-growing areas on infested tubers. Although cyst nematode does not occur in the western states, potato growers and pest control advisers should be prepared to recognize its symptoms in case it is introduced.

If cyst nematode is introduced, prompt control action may prevent it from becoming established or spreading. Once established, this nematode is almost impossible to eradicate, because it forms egg-containing cysts that can survive for at least 20 years in the absence of a host.

Weeds

Weeds reduce yields by competing with potatoes for light, water, and nutrients. Weeds may also be hosts for insects, pathogens, or nematodes that attack potatoes. Dodder, a parasitic weed, injures potato plants directly, and the rhizomes of nutsedge and quackgrass may penetrate tubers, causing losses in quality. Heavy weed infestations reduce harvesting efficiency, increase mechanical damage to tubers, and increase the number of tubers left in the ground or lost over deviner chains.

Management Guidelines

An effective weed management program takes into account the primary weed problems, crop rotation, cultivation, available herbicides, and the competitive ability of the potato crop. Competition from early season weeds will reduce yields if they are not controlled within 4 to 6 weeks after potatoes emerge. Weeds that emerge after vines have covered the rows usually will not compete with the potato crop; however they may reduce yields by interfering with harvest and can produce seed that will cause infestation of subsequent crops. Weeds frequently become more serious if potatoes are set back by adverse conditions early in the season. Weed problems can be reduced by establishing a vigorous stand of potatoes. Prepare fields properly, plant good-quality seed tubers, use spacings that will give an adequate plant stand, maintain adequate soil moisture and fertility, and control diseases, insects, and nematodes.

Field Selection and Crop Rotation

Efficient weed management requires matching the right crop with the right field. You must know which weeds will be present if you are to make this choice. Survey the weeds in the field during the previous crop, then check the herbicide susceptibility chart (Figure 39) to see if herbicides will control the major weeds present. Be sure that conditions will permit proper application of the necessary herbicides. Careful field preparation and irrigation control are required for herbicides that are applied

	HERBICIDES							
	SOIL-APPLIED					FOLIAR-APPLIED		
	EPTC	METOLACHLOR	METRIBUZIN[1]	PENDIMETHALIN	TRIFLURALIN	DINOSEB[2]	GLYPHOSATE[3]	PARAQUAT[4]
PERENNIALS								
Canada Thistle	No useful control	No useful control	Partial control	No useful control	No useful control	No useful control	Partial control	No useful control
Field Bindweed	No useful control	No useful control	No useful control	No useful control	No useful control	No useful control	Partial control	No useful control
Purple Nutsedge	Partial control	●	No useful control	No useful control	No useful control	No useful control	Partial control	No useful control
Quackgrass	Partial control	No useful control	Partial control	No useful control	No useful control	No useful control	Control	No useful control
Yellow Nutsedge	Partial control	Partial control	Partial control	No useful control	No useful control	No useful control	Partial control	No useful control
PARASITIC								
Dodder	No useful control	No useful control	No useful control	No useful control	No useful control	Control	Control	No useful control
BROADLEAVED ANNUALS								
Black Nightshade	Control	+	No useful control	No useful control	No useful control	Control	Control	No useful control
Burning Nettle	Partial control	Control	Control	Partial control	Partial control	Partial control	No useful control	Partial control
Cutleaf Nightshade	Control	+	Partial control	Partial control	No useful control	Control	Control	Control
Groundcherries	No useful control	Control	No useful control	No useful control	No useful control	Control	Partial control	No useful control
Hairy Nightshade	+	+	Partial control	No useful control	No useful control	Control	Control	Control
Common Cocklebur	No useful control	No useful control	Partial control	No useful control	No useful control	Control	Control	Control
Kochia	Partial control	Partial control	Control	Partial control	Partial control	Control	Control	Control
Lambsquarters	+	Partial control	Control	+	+	Control	Control	Partial control
Mallow	No useful control	Partial control	Control	No useful control	No useful control	No useful control	Partial control	No useful control
London Rocket	Control	Partial control	Control	No useful control	No useful control	Control	Control	Control
Mustards	Partial control	Partial control	Control	Partial control	Partial control	Control	Control	Control
Nettleleaf Goosefoot	Control	Partial control	Control	Control	Control	Control	Control	Control
Pigweeds	Control	Control	Control	Control	Control	Control	Control	Control
Prostrate Knotweed	Partial control	No useful control	No useful control	Control	Control	Partial control	Control	Partial control
Common Purslane	Partial control	Control	Partial control	Control	Control	Partial control	Control	Control
Russian Thistle	●	Partial control	Partial control	Partial control	Partial control	Partial control	Control	Control
Shepherdspurse	Partial control	No useful control	Control	No useful control	No useful control	Control	Control	Control
Annual Sowthistle	Control	Control	Control	No useful control	No useful control	Control	Control	+
Sunflower	●	●	Partial control	No useful control	No useful control	Control	Control	+
Wild Buckwheat	●	Partial control	+	+	Control	Control	Control	Control
ANNUAL GRASSES								
Barnyardgrass	Control	Control	Partial control	Control	Control	●	Control	Partial control
Foxtails	Control	Control	Partial control	Control	Control	Partial control	Control	Control
Longspine Sandbur	Control	Control	●	Control	Control	Control	Control	Control
Volunteer Grains	Control	Partial control	Partial control	Partial control	Partial control	●	Control	Control
Wild Oats	Control	Partial control	Partial control	Partial control	Partial control	No useful control	Control	+
Witch Grass	Control	Control	Partial control	Control	Control	Partial control	Control	Control

[1]Metribuzin may also be foliar-applied to cultivars that are not sensitive.

[2]Dinoseb is a restricted-use pesticide in California and Oregon; a permit is required for purchase or use.

[3]Must be applied before potatoes emerge if applied in potato fields.

[4]Paraquat must be applied before planting or as a vine-killing agent. Paraquat is a restricted-use pesticide in all states; an applicator's license is required. In California a permit is required for purchase or use.

Key:
- Control.
- + Control or partial control.
- Partial control.
- ● Partial control or no useful control in some situations.
- No useful control.

Figure 39. Susceptibility of common weed species to major herbicides used in potatoes in 1986.

REGISTRATION STATUS OF THESE COMPOUNDS MAY CHANGE. CHECK WITH LOCAL AUTHORITIES BEFORE USING.

through sprinklers and for some others. Some herbicides may cause injury if applied under poor growing conditions. Fields with major variations in soil type may require changes in herbicide rates or avoidance of some herbicides.

Crop rotations help control difficult weed problems because they allow a variety of weed control methods. Different planting times and cultural practices required for rotation crops may eliminate conditions favoring particular weeds; different herbicides and tillage practices are usually available in alternate crops. Choose rotation crops that compete well with problem weeds and allow effective alternative herbicides, then follow active weed control programs in those crops. Grains are valuable in rotations; herbicides that cannot be used in potatoes can be used for controlling problem weeds in the crop and after harvest. Corn competes well with weeds, tolerates a variety of herbicides, and may also help reduce harmful nematode populations. Control of troublesome perennial weeds may be achieved by discing and watering after a grain harvest to encourage weed growth in the fall, then applying a translocated (systemic) herbicide to regrowth of the perennials. Cotton rotations allow alternative herbicides. Alfalfa may be a useful rotation crop because it shades out a number of annual weeds common in potatoes, and repeated mowing of alfalfa will help reduce infestations of some troublesome weeds. However, the susceptibility of alfalfa and cotton to some root-knot nematodes may diminish their usefulness. Other potato family crops such as tomato, pepper, and eggplant are not good in rotations because the available herbicides are similar to those used in potatoes.

Check with your local extension agent, farm advisor, or other experts to find out which herbicides are registered for use in the rotation crops suited to your area, then choose a combination of crop and herbicide that will control the weeds present. Check the labels to make sure herbicide residues will not damage subsequent crops. Avoid repeated applications of the same or similar herbicides that will encourage buildup of weeds not controlled by those herbicides and the possible buildup of herbicide-degrading microbes in the soil.

Monitoring

To plan a weed management program, you must know what kinds of weeds are present, which ones are most abundant, and whether their abundance is changing. Most of the herbicides used in potatoes are effective only on germinating or very young weeds, so it is essential to know what the target weeds are before they emerge. Normally this information comes from routine weed surveys made in previous seasons. Keep separate records for all fields; this will help you select the correct controls

and detect changes in weed populations that may require special attention.

Monitoring is useful for deciding if additional herbicide applications are necessary and may help reduce the amount of herbicide needed for weed control. After applying an herbicide such as EPTC, which is easily leached from the soil, watch for weed emergence in areas such as the centers of center-pivot sprinklers or along the lateral lines of solid set sprinklers, where most leaching will occur. If using metribuzin, you can reduce the likelihood of injury by using lower rates, then monitoring your fields and applying metribuzin again if weeds emerge.

Surveys. Regularly survey your fields to look for changes in weed populations and to check the effectiveness of weed control measures. If possible, make your first survey while the previous crop is still in the ground. Repeat surveys during each growing season to check on the effectiveness of weed control measures. Make surveys once each week during the early part of the potato season.

To make a weed survey, walk through the field in a random pattern and rate the degree of infestation for each weed species. Check fencerows and ditchbanks as well as the field itself. Pay special attention to perennials and dodder infestations; sketch a map of the field and mark where they occur. Maps will allow you to recheck these areas each season and apply special controls during fallow or rotations.

Weed survey forms are helpful for keeping records; a sample is shown in Figure 40. Whether you use a printed form or not, be sure to take written notes in the field and keep them as part of the permanent field history. Weed survey information collected over several years is extremely valuable in identifying changes in weed populations and in planning herbicide and rotation programs.

Soil Tests. If the field has been plowed before you can do a weed survey, you can use a simple soil test to identify at least some of the weeds present. The test does not work for certain weeds, such as dodder and shepherdspurse, and results are less reliable than those of a survey, but they are better than no information.

Collect samples of the top 2 to 3 inches of soil at random from at least 10 to 20 places in each field. Mix the samples from each area and put the soil in a greenhouse flat or similar container. Moisten the soil with a solution of 100 parts per million (0.01%) gibberellic acid (available commercially) and keep it indoors where it is warm and sunny. Use the photos in this manual and in the weed identification handbooks listed in the References to identify weed seedlings that emerge. Call your local extension agent, farm advisor, or other experts if you need help.

WEED INFESTATION RECORD

FIELD LOCATION _____   CROP _____

DATE NOTES TAKEN _____   DATE PLANTED _____

PREVIOUS CROP _____   HERBICIDE/DATE APPLIED _____

COMMENTS _____   HERBICIDES IN PREVIOUS CROP _____

☐ Dodder

PERENNIALS

☐ Canada Thistle
☐ Field Bindweed
☐ Nutsedge, Purple
☐ Nutsedge, Yellow
☐ Quackgrass
☐ _____

BROADLEAVED ANNUALS

☐ Black Nightshade
☐ Cutleaf Nightshade
☐ Hairy Nightshade
☐ Groudcherries
☐ Burning Nettle
☐ Cocklebur
☐ Field Pennycress
☐ Knotweeds
☐ Kochia
☐ Lambsquarters
☐ London Rocket
☐ Mallow

☐ Mustards _____
☐ Nettleleaf Goosefoot
☐ Pigweed
☐ Purslane
☐ Russian Thistle
☐ Shepherdspurse
☐ Sowthistle
☐ Sunflower
☐ Tansymustard
☐ Tumble Mustard
☐ Wild Buckwheat
☐ _____

ANNUAL GRASSES

☐ Barnyardgrass
☐ Foxtail
☐ Lovegrass
☐ Sandbur
☐ Volunteer Grains _____
☐ Wild Oats
☐ Witchgrass
☐ _____

Figure 40. Example of a weed monitoring sheet. To indicate the level of infestation, use a scale from 1 to 5: 1 = very few weeds; 2 = light infestation; 3 = moderate infestation; 4 = heavy infestation; 5 = very heavy infestation. Alternatively, you can use L to H: L = light, M = moderate, H = heavy.

Herbicide Residue Test. Certain herbicides used in rotation crops may leave residues harmful to potatoes; these include atrazine, chlorsulfuron, diuron, hexazinone, picloram, simazine, and terbacil. If potentially harmful herbicides have been used, you may want to test the soil to see if residues have dissipated before planting a potato crop. Take samples from several different places in the field. At each sample location, dig soil down at least 1 foot, mix it thoroughly, and take enough to fill each of three to five soil buckets or pots with 5 to 10 pounds of soil. As a check, take at least one sample from a location you are sure is free of herbicide residues. Plant one or two seed pieces in each of the buckets or pots, apply some starter fertilizer, and grow the plants until they are at least 4 inches high. Look for injury symptoms by comparing the test samples with the checks. (See *Herbicide Injury* in the *Physiological Disorders* chapter for examples of herbicide injury symptoms.)

Cultivation

During field preparation and hilling, cultivation is an effective way to manage annual weeds early in the season. Cultivation is often the only management available for perennial weeds after potatoes have emerged; it is essential for managing annual weeds that are not controlled by available herbicides. Cultivations during hilling operations enhance weed control by uprooting seedlings. Use a rolling cultivator behind hilling blades to increase the uprooting of early emerging weeds during hilling operations. Additional hilling may be necessary for cultivars such as Kennebec that have shallow tubers. The weed control provided by cultivations may eliminate the need for herbicide applications in some situations. However, the general recommendation for Russet Burbank is to keep cultivation and hilling operations to a minimum.

Additional cultivation can be used just before tuber initiation begins—usually 30 to 40 days after emergence—to control later-emerging weeds. If cultivation controls weeds until potato vines cover the ground, herbicides may not be needed. However, solid set sprinklers may restrict the number of cultivations possible in some areas, such as the San Joaquin Valley of California. Cultivations in addition to normal hilling generally are not recommended in most areas because they can compact soil, damage potato roots and shoots, cause loss of soil moisture, and spread pathogens. In field-stored potatoes, additional weed control is obtained by hilling after vines are killed.

Herbicides

Soil-applied herbicides usually are mixed mechanically or irrigated into the soil before weeds emerge and kill germinating weed seeds or emerging weed seedlings. *Foliar-applied* herbicides are sprayed onto the foliage of weeds after they have emerged, usually while they are small and actively growing. Glyphosate is a *translocated* herbicide; when applied to the foliage it moves throughout the plant, killing both aboveground and belowground parts. Translocated herbicides are most effective when applied to new growth. When used in potato fields, glyphosate must be applied before potato emergence.

With the exception of dinoseb, glyphosate, and paraquat, most herbicides used in potatoes are selective, killing some weeds but not others; most kill only germinating or newly emerged weeds. Choose herbicides that will control the important weed species in your field, and plan to kill weeds before they grow beyond the seedling stage.

Depending on the herbicide used and the weeds to be controlled, the time of herbicide application may be *preplant*, *preemergence* (before potatoes emerge), or *postemergence* (after potatoes emerge). Special precautions may be necessary with preemergence or postemergence application of some herbicides to avoid injuring potato sprouts or vines.

Details of a specific weed control program depend on weed species present, soil type, cultural practices, cultivar grown, and available herbicides. Two or three herbicide applications usually are made during the potato season. A preplant, soil-applied herbicide is mixed into the soil at the same time as field preparation and broadcast fertilizer application. Depending on the weed problems, preemergence herbicides may be applied in some areas 3 to 4 weeks after the preplant application. Usually a combination of two soil-applied herbicides is mixed into the soil during or just after hilling. If additional control is necessary after potatoes emerge, certain herbicides may be applied through sprinkler irrigation. Vine-killing agents are usually contact herbicides. Hilling of field-stored potatoes helps control weeds after vines are dead, and contact herbicides may be used if additional control is necessary.

Determine exactly how the herbicides you choose are to be applied and plan every detail into your cultural program. Some can be applied through overhead sprinklers; this requires a well-designed system with accurate metering, flow-controlled nozzles where land is not level, and check valves to prevent contamination of the water supply. Check with your local extension agent, farm advisor, or other experts before applying herbicides in this way. Never apply herbicides by sprinkler irrigation if the wind speed is over 10 miles per hour. The herbicide EPTC can also be applied in furrow irrigation water.

Proper herbicide rates depend on the types of weeds to be controlled and on the type of soil in your field. Higher rates of some herbicides are required on soils high in organic matter. Consult your local extension agent, farm advisor, or other experts for the rates that work best

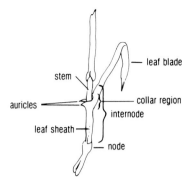

GRASS STEM AND LEAF

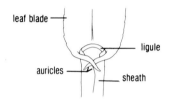

COLLAR REGION OF GRASS

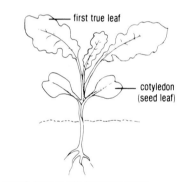

BROADLEAF SEEDLING

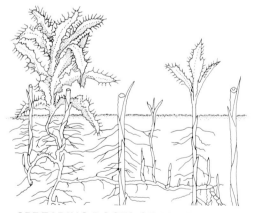

SPREADING ROOTS OF PERENNIALS

Figure 41. Vegetative parts commonly used in identifying weeds.

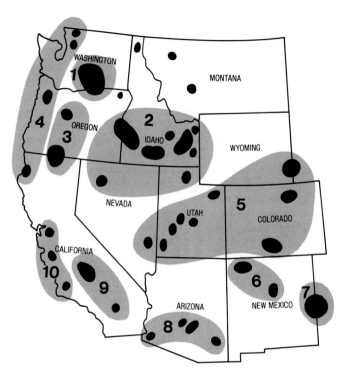

Figure 42. Potato-growing areas of the western states that have similar weed-emergence dates. Estimates of weed emergence for these areas are listed in Table 15.

in your area. Some herbicides may damage certain cultivars. If you do not know the sensitivity of a cultivar, test the application on a few plants first. Find out if subsequent crops will be affected by the herbicides you use and if special efforts such as deep plowing will be needed to protect them from herbicide residues. *Always follow label directions carefully*, and be aware that registrations and restrictions change.

Major Weed Species in Potatoes

The weeds that infest potatoes can be grouped biologically into four categories, each of which may require special control practices: *perennial weeds*, which have vegetative parts that survive from year to year; one *parasitic weed* (dodder) that feeds directly on potato plants; *annual broadleaved weeds*; and *annual grasses*. Vegetative plant parts used in identifying weed species are illustrated in Figure 41. Geographic location, growing conditions, and cultural practices determine which weed species may be problems in a particular growing area. Weeds that occur in potatoes to a significant extent in one or more of the potato-growing areas of the western states are discussed here. Emergence dates of weed species important in the growing areas identified in Figure 42 are listed in Table 15.

Table 15. Weed Emergence Dates for Potato-growing Areas in the Western States.

Weed Species	Weed Emergence Dates in Growing Area (see figure 42)									
	1	2	3	4	5	6	7	8	9	10
Barnyardgrass	Apr–May	May–Aug	May	Apr–Jun	Jun–Jul	Apr–Aug	Apr–Aug	Feb–Mar	Mar	May
Canada Thistle	Mar–f[a]	Mar–Oct	May	Mar–f	Apr–May	—[b]	Mar–f	—	—	—
Cocklebur	May	May–Aug	May	—	May–Jun	May–f	May–f	Mar–Apr	Mar–Apr	Jul
Dodder	Apr–f	May–Jun	Jun	Apr–f	Apr–May	May–f	—	Apr–May	—	May
Field Bindweed	Mar–f	Apr–Sep	May	Mar–f	May–Jul	Apr–f	Apr–f	Mar	Mar	May
Foxtails	May–f	May–Aug	May	May–f	Apr–Jun	Mar–f	Apr–f	—	Mar	Mar
Groundcherries	—	May–Aug	—	—	—	Jun–f	—	Mar–Apr	May	May
Kochia	Mar–f	Apr–Aug	Jun	—	Apr–May	Mar–f	Mar–f	—	—	—
Lambsquarters	Mar–f	Apr–Aug	Apr	Apr–f	May–Jun	May–Aug	Mar–f	Sep–Oct	Feb	May
London Rocket	—	Aug–Mar	Apr	—	—	—	—	Oct	Aug–Mar	Apr
Mallow	Mar	Apr–May	May	Mar–Apr	Apr–May	May–f	—	Oct	Sep–Feb	Mar
Mustards	all year	Aug–May	Apr	all year	May–Jun	Jan–Mar	Jan–Mar	Oct	Aug–Mar	Mar
Nettle	—	—	—	—	—	Jun–Jul	—	—	Nov–Feb	all year
Nightshades	Apr–f	Apr–Sep	May	Apr–f	Jun–Jul	May–f	—	—	Feb–Aug	May
Nutsedge	Apr	Apr–Sep	—	Apr–May	—	—	—	Mar–Apr	Mar	May
Pigweeds	Apr	Apr–Aug	Apr	May	May–Jun	Mar–Aug	Mar	Feb–Mar	Mar–Apr	Apr
Purslane	May–Jul	Jun–Jul	Jun	Apr–Jul	Jun–Jul	Apr–Jul	—	Mar–Apr	Apr–May	May
Quackgrass	Mar–f	Mar–Oct	Apr	all year	Apr–May	—	May	—	—	—
Russian Thistle	Feb–Jul	May–Aug	Jun	—	May–Jun	Apr–Jul	May	Mar,Sep	Feb	Apr
Sandbur	May	May	—	—	Apr–Jun	Jun–Aug	May	Feb–Mar	—	—
Shepherdspurse	Aug–Sep	Sep–May	Apr	Sep–Jun	Jun	Feb–Mar	—	Oct	Aug–Mar	Mar
Sowthistle	Apr	Sep–Jun	May	all year	May	—	—	Oct	all year	Mar
Sunflower	May–Sep	May–Aug	—	—	Jun	May–Jun	May–Jun	Mar–Apr	—	—
Wild Oats	Mar	Mar–Jul	Apr	Sep–Jun	Apr–May	—	—	Oct	—	Mar
Witchgrass	May–Aug	May–Aug	—	May–Aug	Jun	—	—	—	—	—

a. f = frost
b. Dash indicates weed is not important in the area.

Perennial Weeds

Perennial weeds are often the most difficult to control because they have persistent rhizomes, roots, or tubers that survive when aboveground parts of the plants are killed. Avoid fields with heavy infestations whenever possible. Control perennials before planting potatoes; once they establish in a growing crop, little can be done to control them except roguing and cultivating. Control seedlings whenever possible to prevent plants from getting established. The basic strategy for established plants is to destroy as much of the vegetation as possible, then prevent regrowth of the green parts. Useful methods include planting rotation crops in which alternative cultural practices and herbicides can be used, use of herbicides during the season that prevent or retard top growth of perennials, use of foliar-applied, translocated herbicides when crops are not present, and fumigation. If growth of green vegetative parts can be prevented, the energy stored in underground parts will eventually be exhausted and the infestation will die out. Serious perennial weed infestations will require persistent control programs for several years.

Apply a translocated herbicide such as 2,4-D to new weed foliage during a rotation, or apply glyphosate after harvest. Apply herbicides to new leaves that are actively growing but not under stress. In alfalfa rotations, apply herbicides 3 weeks after mowing.

Soil fumigation will control quackgrass and partially control other perennial weeds. Field bindweed and

A

B

occurs, because its extensive, perennial root system makes it difficult to control. Canada thistle can be a problem in all potato-growing areas except Arizona, central and southern California, and central and northwestern New Mexico.

Seed production by Canada thistle is sometimes limited because male and female flowers are formed on separate plants, and entire patches of Canada thistle may be only male or female. Seeds are spread by wind and may be transported as a contaminant in commercial seed; when buried they remain viable for many years. Seedlings begin forming perennial rootstock when they are 3 weeks old or have five leaves. The roots act as storage organs; if chopped up during cultivation, each piece can regenerate a new plant. Shoots from Canada thistle roots begin generating food reserves immediately in the summer or fall; in the spring they generate food reserves when about 1 foot (30 cm) tall.

Canada thistle control requires the use of cultivation and herbicides in both potatoes and rotation crops. The objective is to set back top growth before it can regenerate food reserves or produce seed. Sanitation is needed to prevent spread to new areas and to eliminate the weed from noncrop locations. Most herbicides that can be used in potatoes do not control Canada thistle. Metribuzin will control top growth; apply it just after thistles emerge and repeat the application at least 10 days to 2 weeks later but before vines close. Single applications of glyphosate are effective if applied when thistle plants are at late bud or early bloom stage. Glyphosate is also effective if applied to small Canada thistle plants in the fall, when soil moisture is adequate. Use glyphosate in noncrop areas or before crop plants emerge.

Mowing alfalfa hay or using dicamba* and 2,4-D* in grain will reduce Canada thistle infestations. Plan rotations to allow time for a fall herbicide treatment before the final killing frost. After mowing alfalfa or discing grain stubble, irrigate to encourage regrowth of the thistles. Wait at least 3 weeks or, if possible, until after the first light frost, then apply a translocated herbicide; this treatment is effective as long as thistle plants are still green, even after several hard frosts. If potatoes are to be planted the following spring, avoid using high rates of dicamba, which may leave harmful residues. This program will also control field bindweed infestations.

A. Canada thistle flowers form purple heads about 1/2 inch (1 cm) in diameter.

B. Canada thistle root systems may be 20 feet (6 m) deep and extend 15 feet (4 1/2 m) horizontally.

Canada thistle are usually only set back by fumigation because their perennial roots penetrate so deeply. Generally, soil fumigation is economical for weed control only in small, heavily infested areas, or where it is also necessary to control nematodes or Verticillium wilt.

Because rhizomes, roots, and tubers of perennial weeds are readily moved with soil and water, sanitation is essential to prevent infestations from spreading. Destroy isolated infestations by hand or with herbicides. Avoid cultivating infested fields when the soil is moist unless herbicides are to be used; otherwise, large numbers of new plants may start from chopped up rhizomes or roots. Clean equipment to prevent moving infested soil, and put screens in irrigation ditches to prevent moving viable plant parts in water.

Canada Thistle   *Cirsium arvense*

Found in most of the western states, Canada thistle has the potential to become a serious weed pest wherever it

*Restricted-use pesticide in California; permit required from county agricultural commissioner for purchase or use.

Quackgrass *Agropyron repens*

Quackgrass, also known as couchgrass or wiregrass, is found throughout the northern half of the United States. Quackgrass rhizomes may penetrate potato tubers, reducing their quality. If rhizomes are chopped up, each piece can give rise to a new plant; quackgrass rhizomes are easily spread from field to field on contaminated cultivation equipment. Plants produced from rhizomes emerge 1 to 3 weeks earlier than seedlings. New plants begin producing rhizomes when they have three or four leaves.

Quackgrass can be controlled, but not eradicated, with preplant application of EPTC. Work the area to be treated thoroughly to cut quackgrass rhizomes into small pieces before applying the herbicide. Mix EPTC into the soil by discing 6 inches (15 cm) deep in two directions. This usually controls quackgrass for one season and reduces the rhizomes. Fall treatments with EPTC may provide additional control. Preplant applications of dalapon will reduce the amount and spread of quackgrass but, like EPTC, will not eradicate the weed. Treat actively growing plants when they are 4 to 6 inches (10 to 15 cm) tall and wait 7 to 10 days before preparing seed beds. Do not use dalapon where White Rose or red-skinned varieties will be planted or during periods of frost or dryness.

You can use glyphosate to control quackgrass before potatoes are planted or in noncrop areas. Apply glyphosate to actively growing quackgrass that is at least 8 to 10 inches (20 to 25 cm) tall. For complete control, repeated applications may be necessary. Wait at least 5 to 7 days before preparing seedbeds.

Fumigation with dichloropropene* or metham controls quackgrass but is usually economical only for small infestations, if the quackgrass infestation is severe, or if it is necessary to control nematodes or Verticillium wilt.

Cultivation and herbicide applications may need to be repeated to eliminate quackgrass problems. Observe strict sanitation to avoid spreading or reintroducing quackgrass on contaminated equipment or in irrigation water.

C. Quackgrass rhizomes are straw colored, with a scaly appearance and sharply pointed ends that may penetrate potato tubers. Most rhizomes are in the upper 6 inches (15 cm) of soil; they extend up to 8 inches (20 cm) deep and 3 to 5 feet (1 to 1.5 m) laterally.

D. The collar region at the base of each quackgrass leaf has a pair of auricles and a ragged fringe.

E. Quackgrass seeds are formed in a head that resembles a slender, flattened head of wheat. Seeds are not a major means of quackgrass survival.

---

*Restricted-use pesticide in California; permit required from county agricultural commissioner for purchase or use.

C

D

E

Nutsedges *Cyperus* spp.

Yellow nutsedge, *Cyperus esculentus*, also called yellow nutgrass, is a serious weed pest of potatoes in the San Joaquin Valley of California, where it is widespread in other crops such as cotton and tomatoes. Nutsedge has the potential to become serious in areas of Arizona, Idaho, New Mexico, Oregon, and Washington. It can rapidly spread to new areas because its rhizomes and tubers or "nutlets" are easily carried in soil on farm equipment. Nutsedge rhizomes may penetrate potato tubers, particularly when soil moisture is low. Greatest damage is likely if nutsedge growth occurs between vinekill and harvest.

Nutsedges can reproduce from seed, but tubers are the main means of reproduction and spread. Plants produce tubers on rhizomes as deep as 8 inches (20 cm); the tubers remain viable for several years in dry soil. When conditions are favorable, leaves sprout from tubers, the tubers grow a new system of roots and rhizomes, and new rhizomes will in turn produce new tubers. If new sprouts are killed back, a tuber can sprout up to five more times before exhausting its energy reserves. Long-term control of nutsedge requires preventing seedling plants from growing beyond the 5- to 6-leaf stage, when they start producing new tubers.

Yellow nutsedge does not tolerate shade; once potato vines have closed over, further nutsedge growth is usually suppressed. In early season areas, herbicides may not be necessary if potato vines cover the ground before nutsedges begin to emerge. Where nutsedges emerge before vines close, use a preemergence application of metolachlor (not registered for use in Kern County, CA) or EPTC to prevent nutsedge growth until vines have closed; if necessary, apply EPTC again after about 1 month.

Thoroughly apply vine-killing agents to suppress nutsedge growth after vine death and to help prevent tuber damage. Maintaining soil moisture after vine death may also help prevent damage to potato tubers by nutsedge rhizomes. Several herbicides can be used to control or suppress nutsedges in rotation crops; some very effective herbicides are available for grains or alfalfa. However, it may not be possible to eradicate nutsedge even when weed control is used during fallow periods.

Purple nutsedge, or purple nutgrass, *C. rotundus*, is found in the southern areas where yellow nutsedge occurs. It is not as widespread as yellow nutsedge and tends to be localized in wet areas of fields. Purple nutsedge tubers are more susceptible to drying than those of yellow nutsedge; repeated cultivation of dry soil can reduce or eliminate infestations. Purple nutsedge is less susceptible to metolachlor than is yellow nutsedge.

F

G

H

F. Young nutsedge plants are grasslike, but leaves are thicker and stiffer than most grasses. The leaves are V shaped in cross section and arranged in a spiral at the base.

G. Nutsedge grows to a height of 1 to 2 feet (30 to 60 cm). Flowering stalks are triangular in cross section and are usually no longer than the basal leaves in yellow nutsedge (*shown here*). The flowering stalks of purple nutsedge are usually longer than the basal leaves.

H. Tubers of yellow nutsedge are smooth, about 3/4 inch (1.8 cm) in diameter, and have a pleasant, almondlike flavor. They are formed singly on rhizomes, and young tubers have loose outer scales that later drop off. The tubers of purple nutsedge (*not shown*) are formed in chains along rhizomes. They are scaly and reddish, and have a bitter flavor.

Field Bindweed *Convolvulus arvensis*

Field bindweed, also called wild or perennial morning-glory, is severe in some potato fields in several of the western states. Heavy infestations smother potato plants and interfere with harvest. It is a serious weed pest of potatoes in the Clovis, New Mexico area.

Field bindweed spreads from an extensive root system as well as from seeds. Roots may penetrate to a depth of 25 feet (7.6 m) or more and extend laterally several feet. When roots are chopped up by cultivation in moist soil, the individual pieces can generate new plants. Follow careful sanitation procedures to prevent moving bindweed on contaminated cultivation equipment. Field bindweed seeds remain viable for many years, some germinating to produce new seedlings each year. Preplant or preemergence applications of pendimethalin or trifluralin prevent bindweed seedlings from becoming established during the potato growing season. Once bindweed infestations are established, they are extremely difficult to eliminate.

Some foliar-applied, translocated herbicides are useful for control of field bindweed, although they may not kill plants in a single application. For best results with glyphosate, disc the infested area in the fall 4 to 8 weeks before applying the herbicide. Irrigate to stimulate regrowth of the bindweed, and follow with the glyphosate application. Disc the field again after 1 to 3 weeks. If applying glyphosate to bindweed in the summer, treat when flowers are present on more than half the length of the stem. Glyphosate can be applied before potatoes emerge and can be repeated in the fall after harvest; it can also be applied in the fall following a grain rotation when potatoes will be planted in the spring. Two seasons of grain, in which a mixture of dicamba* plus 2,4-D* is used, followed by fall applications of glyphosate as described for Canada thistle control, will set back field bindweed enough to allow a season of potatoes. However, there is a potential for injury to certain varieties when dicamba is used on sandy soils the year before planting potatoes. Repeated treatments over several seasons are necessary to control bindweed.

I. Field bindweed seedlings have seed leaves that are nearly square, with a notch at the tip. Petioles are flattened and grooved on the upper surface. Plants sprouting from roots (*not shown*) lack seed leaves.

J. Prominent white flowers make field bindweed easy to identify. Severe infestations can make harvesting difficult.

---

*Restricted-use pesticide in California; permit required from county agricultural commissioner for purchase or use.

I

J

Parasitic Weed

Dodder *Cuscuta* spp.

A parasite that attacks many crops, weeds, and native plants, dodder is a problem in the lower Snake River Valley and in some potato fields of the Columbia Basin and Klamath Basin, but is not a widespread pest of potatoes. However, its wide host range and the long life of its dormant seeds make dodder hard to control and nearly impossible to eradicate. Dodder is particularly troublesome where seed alfalfa is grown, because the dodder seeds contaminate and ruin the seed crop.

Dodder lacks chlorophyll and can grow only by penetrating host plant tissues to obtain water and nutrients. Seedlings must attach to a suitable host within a few days of germination or they die. Young seedlings twine around the host plant and produce suckers, called haustoria, that penetrate host tissues. After attachment, the part of the dodder stem between the host and soil

K

L

Germinating seedlings can be killed with applications of DCPA (not registered in California); apply this herbicide as soon as dodder seedlings begin to emerge, then irrigate lightly. If infestations are not found until the dodder is established in the vines, destroy the infested plants. If only a few plants are infested, remove them from the field and burn them. Spray larger areas with a contact herbicide, then burn the plants after they dry out. Remove or treat plants several feet beyond the apparent edge of the infestation to be sure all dodder is destroyed. Use these control measures in fence rows, ditchbanks, and roadsides as well as in the field. Destroy dodder before it produces seeds; infestations left until harvest will spread seeds through the field on harvesters or other equipment. Revisit treated spots every 2 weeks to kill any survivors or new seedlings, and mark the spots so they can be checked in following seasons.

Rotations of nonhost crops, such as cereals, beans, or corn can reduce dodder populations, but it may be necessary to keep the field in a nonhost crop for several years to have a significant effect.

K. Threadlike, leafless, yellow dodder stems twine around host plants, creating a tangled mat.

L. Dodder flowers and seed capsules are borne in clusters. Each flower is about 1/8 inch (3 mm) long. Destroy dodder infestations before this stage to prevent seed production.

Annual Broadleaved Weeds— Potato Family

Several potato family weeds are common pests in potato fields. Because they are closely related to potatoes, they are hosts for several potato diseases and insect pests, including all potato viruses, tobacco rattle virus, *Verticillium*, late blight, black dot, nematodes, Colorado potato beetle, and green peach aphid.

Potato Volunteers   *Solanum tuberosum*

In rotation crops, field borders, or cull piles, potato volunteers are a serious problem because they can be reservoirs of potato insects and diseases. Volunteers are often a major source of leafroll and other potato viruses and may also be important sources of ring rot, blackleg, or late blight. Populations of Colorado potato beetle or other harmful insects may build up on volunteers.

In winter potato areas, desiccation during summer fallow periods effectively controls volunteers. A hard winter freeze in the absence of snow cover is one of the most effective controls for volunteers. Tubers are protected against mild winters if buried 1 inch (2.5 cm) deep, and against severe winters if buried 4 to 6 inches (10-15 cm) deep.

dies, and a network of threadlike stems winds through the canopy of the original host plant and may also spread through adjacent plants. Each dodder plant produces thousands of hard seeds that can remain dormant in the soil for at least 5 years; some of the seeds may germinate as soon as they fall to the ground if conditions are favorable.

Dodder commonly invades fields from weedy or natural areas, but seeds can also be introduced in water, in contaminated crop seed, and in the manure of livestock that have eaten infested forage. Dodder is a problem in alfalfa and often appears in potatoes following an infested alfalfa crop or in fields grazed by cattle that have eaten infested alfalfa. Dodder does not always reappear in the same places each year, but once seeds are present, the potential for infestation remains indefinitely. Keep a permanent record of dodder infestations and survey the area regularly every season for reappearance of dodder seedlings.

Most available herbicides do not control potato volunteers. Atrazine can be used in corn, but potatoes may not be planted for at least 1 year after its use. Apply glyphosate or a contact herbicide to volunteers along field borders or to isolated volunteers in rotation crops. Foliar application of a sprout inhibitor greatly reduces volunteers in the following crop. Eliminate cull piles or treat plants that emerge from them with a contact herbicide.

Nightshades *Solanum* spp.

Three nightshade species can be problems in potatoes. The worst of these weeds in most potato-growing areas of the West is hairy nightshade, *Solanum sarrachoides*. Black nightshade, *Solanum nigrum*, is more common in the growing areas of Arizona, central and southern California, and New Mexico. Black and hairy nightshades may be found together; the two species can be distinguished by the shape of the calyx, color or appearance of mature berries, and hairiness of stems and leaves. Cutleaf nightshade, *Solanum triflorum*, occurs sporadically throughout the western states.

The best strategy for controlling nightshades is to plan a crop rotation sequence that prevents populations from building up. Choose alternate crops, such as corn, sorghum, cereals, or sugar beets, that can be managed with herbicides that kill nightshades. Nightshades are prolific seed producers; once a nightshade population builds up, it may take several years' rotations to reduce infestations significantly. Hairy nightshade produces so many seeds that large numbers of seedlings may survive even when an herbicide is 99% effective. Although most seeds germinate after 1 or 2 years, some may survive for up to 30 or 40 years.

Combinations of two or three different herbicides or multiple applications may be necessary to control nightshade infestations in potatoes. Make sure cultivations and herbicide applications will not allow nightshades to develop beyond the seedling stage.

Sanitation is essential to prevent nightshades from building up in fields free of heavy infestations. Destroy nightshade plants before they mature, because seeds from a single plant can produce thousands of seedlings in subsequent seasons. Rogue any plants that have seeded and remove them from the field to prevent spreading the seeds during harvest. Apply a translocated herbicide such as 2,4-D* to nightshades growing along field borders; weeds in field borders are potential sources for field infestations as well as disease and insect problems.

M

N

O

Q

P

* Restricted-use pesticide in California; permit required from county agricultural commissioner for purchase or use.

M. Seed leaves of hairy nightshade are lance shaped; the first true leaves have prominent veins and usually have wavy edges and numerous fine, short hairs, especially along the underside of the main vein.

N. Hairy nightshade berries are green or yellowish brown when mature; occasionally they are brownish black. The calyx covers the entire upper surface of the fruit. The pedicels, like stems and leaves, are usually conspicuously hairy. Mature plants can be up to 2 feet (60 cm) tall.

O. Seed leaves of black nightshade are elongate-oval and pointed; the first true leaves are spade shaped with smooth edges. Lower surfaces are often purple.

P. Black nightshade berries turn from green to black when mature and the calyx covers only a small part of the fruit surface. Petioles, stems, and leaves have some hairs but are not densely hairy or sticky. Black nightshade plants may be erect and bushy, up to 3 feet (1 m) tall, or prostrate and spreading.

Q. Cutleaf nightshade leaves are deeply cleft at the edges. The plant may be erect or prostrate and smooth or somewhat hairy. The berries (*not shown*) are green or yellowish.

Groundcherries *Physalis* spp.

Groundcherries are summer weeds that occur in most areas of the western states, more commonly in the growing areas of southern California and Arizona. Groundcherries are related to nightshades and are also hosts for many potato pests. They are controlled by the same practices used for nightshades.

R. Groundcherry flowers are cup shaped and about 3/8 inch (8 mm) across. All species have a smooth, tomatolike fruit enclosed in a characteristic green, heart-shaped calyx.

R

S

U

T

V

Annual Broadleaved Weeds— Mustard Family

Several common weeds are in the mustard family. They are generally early season or winter annual or biennial weeds, and may grow all year long in cool areas. Mustard family weeds are hosts for tobacco rattle virus, green peach and other aphids, and flea beetles. Green peach aphid and other aphid vectors of potato viruses may build up on these weeds in the spring and then migrate into potato fields later.

Mustard family weeds generally are controlled with normal cultivations and herbicides in potato crops. Use foliar-applied, translocated herbicides to control mustard family weeds in field borders and other noncrop areas, being sure to avoid spray drift.

Mustards *Brassica* spp.

Mustards are common winter or early spring weeds. Some species may grow as biennials. They usually are not serious pests in potatoes, but can become a problem in winter potatoes in the low desert valleys if not controlled before vines close. Mustards are early season hosts for buildup of aphid populations in orchards or field borders.

S. All mustard seedlings in the *Brassica* genus have broad seed leaves with a deep notch at the tip. The first true leaves are bright green on the upper surface and paler below.

T. Mature mustards have dense clusters of yellow flowers at the tips of branches. Leaves are toothed, alternate, and often deeply lobed, especially toward the base of the plant. The species shown here is birdsrape mustard or wild turnip, *Brassica rapa*.

London Rocket *Sisymbrium irio*

A winter annual, london rocket is more common in Arizona and California than in other potato-growing areas of the West. It is a highly competitive weed in grain rotations.

U. The edges of the first true leaves of London rocket seedlings are often somewhat indented. They are very difficult to distinguish from shepherdspurse seedlings. Most or all of the early leaves of London rocket are deeply indented; early leaves of shepherdspurse can be either indented or smooth edged.

V. London rocket flowers form small clusters at the tips of stems that bear long, slender seed pods. The plants usually grow to about 2 feet (60 cm) tall.

Tumble Mustard *Sisymbrium altissimum*

Tumble mustard, also called Jim Hill mustard, is a winter annual weed that is most common in Northwest growing areas (Idaho, Montana, Oregon, Washington, northern Nevada, and northern California). It is frequently a problem in grain rotations. In the Klamath Basin, tumble mustard is the earliest green peach aphid host after peach trees.

W. Tumble mustard plants grow to about 3 1/2 feet (1 m) tall. They branch widely and the upper leaves have a thin, fingerlike appearance. Tumble mustard seedlings are similar to London rocket, but early leaves have conspicuous, coarse hairs.

Shepherdspurse *Capsella bursa-pastoris*

Shepherdspurse is a common weed throughout the western states. The margins of the first true leaves of shepherdspurse seedlings are not indented and are covered with star-shaped hairs that can be seen with a hand lens.

X. Shepherdspurse grows up to 20 inches (50 cm) tall from the center of a rosette of indented and smooth-edged leaves. The heart-shaped seed pods make this species easy to recognize when it is mature.

Field Pennycress *Thlaspi arvense*

A common weed in Northwest growing areas, field pennycress may be a serious pest in grain rotations.

Y. Field pennycress has clusters of small white flowers that branch out from the tips of short plants, about 6 to 15 inches (15 to 38 cm) tall. The seed pods have a flat, palm fan shape, with a notch in the tip.

Tansymustard *Descuraina pinnata*

Tansymustard occurs throughout the western states. It is usually more abundant in the drier growing areas.

Z. Tansymustard is a spring or summer annual that grows up to 2 feet (60 cm) tall, with yellow flowers at the ends of long, slender stems. These stems bear seed pods that are about 1/2 inch (12 mm) long. The seed pods are shorter than their petioles, in contrast to London rocket and tumble mustard. Leaves are finely divided.

X

Y

W

Z

A

B

C

D

Annual Broadleaved Weeds—Goosefoot Family

Weeds of the goosefoot family are common summer annuals in all potato-growing areas of the West. Cultivations and soil-applied herbicides generally control these weeds.

Lambsquarters   *Chenopodium album*

A common weed throughout the western states, lambsquarters is sometimes called white goosefoot or fat-hen. Lambsquarters is a winter weed in the low desert valleys of California and Arizona, where it can be a problem if not controlled before vines close; elsewhere it is a summer weed. Lambsquarters is a host for the beet leafhopper, green peach aphid, and potato virus X.

A. Lambsquarters seed leaves are narrow, with nearly parallel sides. The seed leaves and early true leaves are dull bluish green above and often purple below. The mealy leaf texture distinguishes lambsquarters from most other seedlings.

B. Lambsquarters may grow up to 6 feet (1.8 m) tall, depending on moisture and soil fertility. Tiny, inconspicuous flowers are packed in dense clusters at the tips of the main stem and branches.

Kochia   *Kochia scoparia*

Kochia is a common summer weed in the Rocky Mountain, Snake River Valley, and Klamath Basin growing areas.

C. Kochia seed leaves (*not visible*) are thick, elongated, and often magenta underneath. The first true leaves are grayish, hairy, and borne in tight rosettes.

D. Mature kochia plants are 20 to 60 inches (0.5 to 1.5 m) tall, with a bushy appearance. Small, inconspicuous flowers are formed at the ends of branches. Stems of kochia may turn reddish in the fall.

Russian Thistle   *Salsola iberica*

Russian thistle occurs throughout the western states, usually more abundantly in drier areas. It can be a serious pest of potatoes in the Columbia Basin and Rocky Mountain growing areas. Russian thistle is the preferred host of the beet leafhopper. Russian thistle seedlings closely resemble pine seedlings.

E. Mature Russian thistle plants are spherical bushes up to 5 feet (1.5 m) tall, with small, lance-shaped leaves. When old, the plants turn grayish brown, break off at the soil line, and become "tumbleweeds," spreading their seeds as they are blown about in the wind.

E

F

F. The seed leaves and first true leaves of Russian thistle are long and slender and resemble pine needles.

Annual Broadleaved Weeds— Thistle Family

Weeds in the thistle family may be problems in some potato-growing areas. The perennial Canada thistle is important in many areas. Some thistle family weeds can host certain potato viruses. The weed species in this family vary in their susceptibilities to herbicides used in potatoes and may be difficult to control with chemicals.

Annual Sowthistle *Sonchus oleraceus*

Annual sowthistle is most common in the coastal growing areas of California, Oregon, and Washington, and in the low desert valley growing areas of Arizona and California. It is a winter or early spring weed in most potato-growing areas. It grows year-round in the central coast and San Joaquin Valley growing areas of California, where it can become a problem during the field storage of Kennebecs. Control sowthistle with early season cultivations or applications of EPTC or metolachlor (not registered for use in Kern County, CA) before the weed emerges. Metribuzin is effective on emerged sowthistle, but cannot be used in most areas where this weed is a problem.

G. Seed leaves of annual sowthistle have smooth edges and a grayish powder on the upper surface. First true leaves are broad at the tip and taper at the base. They have slightly wavy edges and may have soft prickles around their margins.

H. Mature annual sowthistle plants are 3 to 6 feet (1 to 2 m) tall. Yellow flowers are borne on the ends of stalks. The base of each long, pointed leaf wraps around the stem. If stems or leaves are cut or torn, a milky juice flows out.

G

H

Common Sunflower   *Helianthus annuus*

Common sunflower is a summer weed that occurs sporadically in Rocky Mountain, Snake River Valley, Columbia Basin, and low desert valley growing areas. It is a serious pest in Colorado, and is a common problem in grain rotations wherever it occurs. Sunflower is a host for potato virus X and tobacco rattle virus. Try to control sunflowers with early season cultivations and preemergence herbicides; they are difficult to control later in the season. Mature sunflower plants grow from 1 1/2 to 9 feet (1/2 to 2 1/2 m) tall and are highly branched. The characteristic yellow flowers make sunflower easy to recognize.

I. Seed leaves of common sunflower are twice as long as they are wide. The first true leaves are similar in shape, with smooth edges and short, rough hairs. Sunflower seedlings have strong taproots.

I

J

K

L

Common Cocklebur   *Xanthium strumarium*

Though a summer weed that occurs throughout the western states, common cocklebur is generally not a serious pest of potatoes. Mature cocklebur plants are bushy, from 2 to 5 feet (1/2 to 1 1/2 m) tall, with heart-shaped leaves and dark red spots or streaks on the stems. The leaves have a rough texture. Female flowers develop into football-shaped burs with stiff, hooked spines. Each bur contains two black seeds. Tobacco rattle virus can be transmitted in cocklebur seeds.

J. The seed leaves of cocklebur are narrow, long, and bright green. The first true leaves are much broader, with indented edges and a dull green upper surface. Cocklebur seedlings are toxic to livestock.

Annual Broadleaved Weeds— Other

Pigweeds   *Amaranthus* spp.

The pigweeds are dominant summer weeds in most potato-growing areas. Redroot pigweed, *Amaranthus retroflexus*, is the most common and is one of the most prevalent weeds in all potato-growing areas. Redroot pigweed is also called rough pigweed, and is a host for the beet leafhopper and potato virus X. Other pigweed species that may occur in potatoes include Powell amaranth (*A. powellii*), tumble pigweed (*A. albus*), prostrate pigweed (*A. blitoides*), and smooth pigweed (*A. hybridus*), which is also called green amaranth. Pigweeds are controlled with cultivations and herbicides commonly used in potatoes; prostrate pigweed is more difficult to control.

K. Seed leaves of redroot pigweed are long, narrow, and bright magenta below. First true leaves have a shallow notch at the tip. Seedlings of other pigweed species are similar.

L. Redroot pigweed may be 7 feet (2 m) tall in some situations. Branches rise mainly from the base; stems are furrowed. Leaves may be several inches long, with long petioles and prominent veins on the lower surface. Flowers are arranged in dense spikes at the tops of the main stem and branches, and in smaller clusters in leaf axils.

Mallow   *Malva* spp.

Little mallow, *Malva parviflora*, occurs in the potato-growing areas of the central California coast and the low desert valleys. It can grow year-round in these areas.

Little mallow is the most serious weed pest of potatoes in the Salinas Valley of California. Common mallow, M. *neglecta*, occurs in other areas, where it is a summer annual or biennial weed. Both species are often called cheeseweed, and they are difficult to distinguish from one another. Cultivate to destroy mallow seedlings while they are small; otherwise they will rapidly form deep, tough taproots. Mallow is difficult to control with available herbicides.

M. Seedlings of little mallow (*shown here*) and common mallow are similar. Seed leaves are triangular or heart shaped. True leaves are round or kidney shaped, with scalloped edges.

N. Common mallow can grow up to 4 feet (125 cm) tall; little mallow (*shown here*) can be prostrate or grow up to 5 feet (150 cm) tall. These weeds are frequently called cheeseweed because the distinctive fruits resemble tiny wheels of cheese.

M

N

Nettle   *Urtica urens*

Burning nettle, also called stinging nettle, is a problem in California, in the San Joaquin Valley where it is a winter weed, and the central coast areas where it can grow year-round. Encourage nettle seedlings to emerge before potatoes by applying a light irrigation if there is no rain, then remove them with cultivation. Use a contact herbicide if cultivations are not possible.

O. Seed leaves of burning nettle are round or slightly elongated with smooth edges and a small notch in the tip. The first true leaves have distinctly toothed margins.

P. Mature nettle plants are 5 to 24 inches (12 to 60 cm) tall, with stems branching at the base. Stems are square in cross section; both leaves and stems have stinging hairs.

O

P

Wild Buckwheat and Prostrate Knotweed
Polygonum convolvulus, P. aviculare

Wild buckwheat and prostrate knotweed are common weeds in many potato-growing areas. They are controlled with cultivation and available herbicides.

Q. True leaves of wild buckwheat are shaped like arrowheads and are more pointed than leaves of field bindweed, with which it is sometimes confused. Mature plants have trailing or twining stems 8 to 40 inches (20 to 100 cm) long. The inconspicuous flowers form clusters at the base of leaf stalks or at the end of stems.

Q

R   S

T   U

Annual Grasses

Many annual grasses, including volunteer grains, are common weeds in potato fields. Most are controlled with available herbicides; repeated applications may be necessary where grasses are prevalent, if shading by potato vines is not sufficient to prevent the weeds from developing and setting seed. Corn, wheat, and barley volunteers are most difficult to control. Grazing livestock on stubble after a corn harvest will reduce the number of corn volunteers.

Barnyardgrass *Echinochloa crus-galli*

Also called watergrass, barnyardgrass is one of the most common summer annual grass pests. It occurs in all potato-growing areas.

T. The best way to recognize barnyardgrass is to strip off leaves and look at the collar region. It is the only common summer weedy grass that lacks a ligule.

U. Barnyardgrass plants can grow up to 6 feet (2 m) tall where water is available in the summer, but may be only 6 inches (15 cm) tall where it is drier. Some plants may root at the lower nodes, forming large clumps. Flower heads of some varieties have long bristles. Each barnyardgrass plant can produce a large quantity of seed.

Foxtails *Setaria* spp.

Summer weeds throughout the western states, foxtails are more common in the northern growing areas. Yellow foxtail (*Setaria glauca*) and green foxtail (*S. viridis*) are distinguished from each other and from other summer grasses by characteristics of the collar and leaf sheath. Foxtails, sometimes called bristlegrasses, are frequent pests of alfalfa.

V. Mature foxtail plants are 1 to 3 feet (30 to 90 cm) tall, with branching and some spreading at their bases. Leaf blades are 4 to 15 inches (6 to 38 cm) long, and most have a spiral twist. Flower heads of yellow foxtail (*shown here*) are dense spikes with yellow or reddish bristles that are 1/4 to 1/3 inch (6 to 8 mm) long. Green foxtail flower spikes have green or purple bristles.

W. The ligule of yellow foxtail is a fringe of hairs and there are no auricles. There are no hairs on the leaf sheath margin below the collar as on green foxtail.

X. If the leaf sheath of green foxtail is pulled away from the stem, you can see fine hairs on the leaf sheath below the collar region. The growth habit of green foxtail is similar to yellow foxtail, but the leaf blades of green foxtail are flat and lack hairs.

Common Purslane *Portulaca oleracea*

Common purslane occurs in all potato-growing areas, and is usually controlled with cultivation and available herbicides.

R. Seed leaves of common purslane are smooth and succulent with a reddish tinge. The first true leaves have rounded tips that are broader than their bases. Newly emerging leaves on young plants are flattened together.

S. Mature purslane plants form mats of highly branched, reddish stems that are 1/2 to 3 feet (15 to 90 cm) long and turned up at their tips. The pale yellow, cup-shaped flowers are usually open just in the morning.

V

W

X

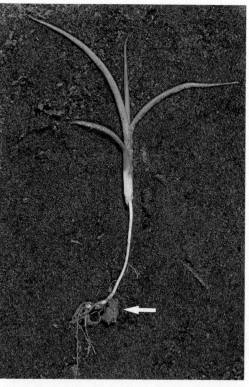

Y

Longspine Sandbur *Cenchrus longispinus*

Sandbur occurs sporadically throughout the western states, growing in sandy soils. It may be a problem in the Columbia Basin, Colorado, New Mexico, and Wyoming.

Y. Sandbur seedlings closely resemble those of barn-yardgrass. However, if you carefully dig up a sand-bur seedling, you can often find the bur still attached (arrow).

Z. Sandbur stems are weak and branched, and the plant often spreads along the ground. Sandbur plants sometimes have a dense, bushy growth. Flower heads consist of several burs (as many as 40) in a spike that is partially enclosed by a leaf sheath. The burs are yellowish green when young and turn a light brown when they mature.

Z

A

B

C

D

Wild Oats *Avena fatua*

A pest throughout the western states, wild oats is most serious in the Northwest growing areas of northern California, Oregon, Washington, and Idaho. Wild oats seeds are a favorite forage of meadow mice; where meadow mice become a problem in potatoes, their burrows are frequently found near wild oats plants.

A. Wild oats plants are 1 to 4 feet (30 to 120 cm) tall, with several stems growing from the base. The flowers are large and open, resembling those of cultivated oats.

B. A large, whitish ligule is present on the collar of a wild oats leaf blade. Few or no hairs are present on the leaf sheath; hairs are generally present on the margins of the leaf blade.

Witchgrass *Panicum capillare*

A summer annual weed, witchgrass is sometimes a problem in potato-growing areas of Colorado, the Columbia Basin, and the Snake River Valley.

C. The mature witchgrass plant is bushy, branched from the base, and has a fuzzy appearance.

D   The stem, sheath, and leaf blade of witchgrass are covered with long, coarse hairs. The ligule consists of a fringe of hairs.

Vertebrates

Vertebrates generally are not significant pests of potatoes in the western United States. In many growing areas, minor tuber damage may be caused every year by pheasants, geese, ground squirrels, or meadow mice, but control measures are not necessary. In some locations, particularly in the Northwest coast areas, pocket gophers may be a problem if they have built up in a rotation crop; controls are generally not applied in potatoes. Controls are required for meadow mice in parts of the Klamath Basin during the occasional years when mouse populations reach damaging levels.

Meadow mice (Figure 43), also called voles or field mice, live in colonies in areas such as irrigated pastures, fencerows, or weedy ditchbanks, where the soil is suitable for burrowing, and where vegetation provides cover. They usually avoid the sandy soils in which potatoes are commonly grown in the western states. The soil of the Tule Lake Basin of northern California is more favorable for meadow mice, and there is also abundant natural habitat where mouse populations can remain active year-round. In this area meadow mouse populations reach levels that require control every 7 to 10 years. Controls are required less frequently in other parts of the Klamath Basin.

Meadow mice move into potato fields when infested grain or alfalfa fields are harvested, usually in August or September, burrowing into hills for shelter and to feed on tubers. Their feeding damages tubers directly and their burrows may expose tubers to sunlight or freezing temperatures, causing additional losses. Damage may also be caused by predators that dig into potato hills in search of mice. If mouse populations are high at harvest, mice will be carried into storage, where they will continue to feed on tubers.

In the Tule Lake Basin, ditchbanks are monitored for meadow mice each spring by the county agricultural commissioners' offices. When trap monitoring or visual estimates of mouse activity indicate damage is likely, poison baits may be applied to ditchbanks in the spring as a preventive measure, or directly to potato fields after vine death. If monitoring programs in your area indicate that meadow mice may become serious, begin checking

Figure 43. The meadow mouse is 4 to 6 inches long when full grown, with a blunt nose, small eyes, and short, furry ears. The tail is less than half as long as the body and is slightly hairy.

your fields for mouse activity when removing irrigation pipe at the end of the season or after nearby grain or alfalfa fields are harvested. Look for active mouse holes—ones with fresh soil around the entrance. Check closely around clumps of wild oats, where meadow mouse burrows are most likely to be found.

In other Klamath Basin growing areas, there are no regular monitoring programs because meadow mice rarely reach damaging levels. However, a routine check for mouse activity near the end of each season is a good way to identify the occasional year when meadow mice may become a problem. If you find meadow mouse activity in your fields at the end of the season, check with your local county agent, farm advisor, or agricultural commissioner's office to see if bait application is recommended.

References

General

Annual Progress Reports. New Mexico State University Agricultural Sciences Center, Farmington, NM 87499.

Annual Reports: California Potato Research Program, California Potato Research Board, 531 N. Alta, Dinuba, CA 93618.

Commercial Potato Production in North America, Potato Association of America Handbook. Potato Association of America, Orono, ME, 1980, 36pp.

Growing Potatoes in Arizona. Publication 115083 (Arizona).

Oregon State Potato Conference and Trade Fair. Proceedings published annually by the Oregon State Potato Commission, Suite 214 Equitable Building, 530 Center Street N.E., Salem, OR 97301.

Potatoes: Influencing Seed Tuber Behavior. Publication PNW 248 (Idaho, Oregon, Washington).

Potato Growers Newsletter. Potato Lab, Montana State University, Bozeman, MT 59717.

Specific Gravity of Potatoes. Current Information Series No. 609 (Idaho).

Sugar Development in Potatoes. Extension Bulletin 0717 (Washington).

Proceedings of the University of Idaho Winter Crop Schools. Published annually (Idaho).

Washington State Potato Conference and Trade Fair. Proceedings published annually by the Washington State Potato Commission, 108 Interlake Road, Moses Lake, WA 98837.

Degree-Days and Weather

Construction of Weather Shelters. Leaflet 2371 (California).

Degree-Days: The Calculation and Use of Heat Units in Pest Management. Leaflet 21373 (California).

Soil and Water

Available Water-holding Capacities of Soils in Southern Idaho. Current Information Series No. 236 (Idaho).

Basic Irrigation Scheduling. Leaflet 21199 (California).

Diagnosing Soil Physical Problems. Leaflet 2664 (California).

Guide to the Interpretation of Soil and Water Tests. Circular 191 (Nevada).

How to Take a Soil Sample and Why. Extension Circular 628 (Oregon).

Interpreting Soil Analyses. Publication GA126 (New Mexico).

Irrigation: When? How Much? How? Publication 115020 (Arizona).

Making and Using Soil Moisture Tensiometers. Publication EM 3078 (Washington).

Measurement of Plant and Soil Water Status. Research Bulletin 484 (Utah).

Measuring Irrigation Water. Leaflet 2956 (California).

Nozzle Management and Leak Prevention for Sprinkler Irrigators. Current Information Series No. 569 (Idaho).

Principles of Soil Sampling for Northwest Agriculture. Western Regional Extension Publication 9.

Questions and Answers about Tensiometers. Leaflet 2264 (California).

Sandy Soil and Soil Compaction. Information Circular 687 (Oregon).

Water Used by Various Crops, Clovis Area. Publication GM218 (New Mexico).

Water Used by Various Crops, Farmington Area. Publication GM219 (New Mexico).

Nutrients

Applying Fertilizer Through Sprinkler Systems. Extension Bulletin 774 (Oregon).

Applying Nitrogen Fertilizers in Irrigation Water. Publication GA125 (New Mexico).

Critical Nutrient Ranges in Northwest Crops. Western Regional Extension Publication 43.

Fertilizer Guide for Irrigated Potatoes (Central Oregon and Klamath Areas). Publication FG 56 (Oregon).

Fertilizer Guide for Irrigated Potatoes (Columbia Basin—Malheur County). Publication FG 57 (Oregon).

Fertilizer Guide for Irrigated Potatoes (Western Oregon—West of Cascades). Publication FG 19 (Oregon).

Fertilizer Guide for New Mexico. Publication C478 (New Mexico).

Fertilizer Guide: Irrigated Potatoes for Central Washington. Publication FG—7 (Washington).

Fertilizer Placement. Current Information Series No. 757 (Idaho).

Fertilizer Placement for Potatoes. Publication XB 735 (Washington).

Guide to Fertilizer Recommendations in Colorado. Publication XCM-37 (Colorado).

Idaho Fertilizer Guide: Potatoes. Current Information Series No. 261 (Idaho).

Interpretation of Soil Test Nitrogen: Irrigated Crops in Central Washington. Publication EM 3076 (Washington).

Nutrient Use by Potato Vines and Tubers. Current Information Series No. 470 (Idaho).

OSU Soil Test for Nitrate Nitrogen. Extension Circular 770 (Oregon).

Phosphorus Fertilization: Broadcast, Banding, and Starter. Publication EM 3295 (Washington).

Phosphorus Fertilization of Irrigated Soils in Central Washington. Extension Bulletin 0602 (Washington).

Sampling for Plant Tissue Analysis. Publication GA123 (New Mexico).

Scheduling Nitrogen Applications for Russet Burbank Potatoes. Current Information Series No. 637 (Idaho).

Soil and Plant Tissue Testing in California. Bulletin 1879 (California).

Seed Certification

California Certified Seed Potatoes. Article 7, California Department of Food and Agriculture, 1220 N Street, Sacramento, CA 95814.

Colorado Rules and Regulations for Certified Seed Potatoes. Potato Certification Service, Department of Horticulture, Colorado State University, Fort Collins, CO 80523.

Idaho Rules of Certification. Idaho Crop Improvement Association, Inc., 1641 South Curtis Road, Boise, ID 83705.

Montana State University Seed Potato Certification Rules and Regulations. Potato Lab, Montana State University, Bozeman, MT 59717.

Oregon Certified Seed Handbook. (Oregon).

Oregon Potato Seed Certification Standards. (Oregon).

Oregon Seed Certification Program. Extension Circular 1090 (Oregon).

Seed Potatoes: Certification Requirements and Standards. Utah Crop Improvement Association, Utah State University, UMC 48, Logan, UT 82322.

Harvest and Storage

Adjust Potato Harvester Speed to Reduce Bruising. Current Information Series No. 263 (Idaho).

Designing Bulk Potato Structures. Publication PNW 236 (Idaho, Oregon, Washington).

Modifications to Potato Harvesting and Handling Equipment That Can Reduce Bruising. Current Information Series No. 330 (Idaho).

Potato Storage and Ventilation. Publication EM 2799 (Washington).

Potato Storage—Construction and Management. Current Information Series No. 297 (Idaho).

Potato Storage Management. Current Information Series No. 349 (Idaho).

Preventing Condensation on the Ceiling of Potato Storages. Current Information Series No. 299 (Idaho).

Reducing Potato Damage During Harvest. Extension Bulletin 0646 (Washington).

Reducing Potato Harvesting Bruise. Extension Bulletin 1080 (Washington).

Pesticide Application and Safety

Application of Agricultural Chemicals in Pressurized Irrigation Systems. Current Information Series No. 673 (Idaho).

Calibrating Chemical Sprayers. Bulletin 848 (Wyoming).

Calibration and Use of Pesticide Application Equipment. Publication 181113 (Arizona).

Calibration of Herbicide Sprayers. Leaflet 2710 (California). Also available in Spanish, as Leaflet 21099 (California).

Calibration of Pesticide Applicators. Publication GM104 (New Mexico).

Colorado Pesticide Guide—Field Crops. Bulletin XCM-45. (Colorado).

Handling Pesticides Safely. Current Information Series No. 653 (Idaho).

How Much Chemical Do You Put in the Tank? Leaflet 2718 (California).

Managing Unwanted Herbicides in the Soil. Extension Bulletin 1059 (Washington).

Pesticide Application and Safety Training Packet. Publication 4070 (California).

Pesticide Applicator Information. Publication 884056 (Arizona).

Pesticide Safety and the Commercial Ground Sprayer. Publication EM 3673 (Washington).

Pesticide Safety Information Series. California Department of Food and Agriculture, 1220 'N' Street, Sacramento, CA 95814. Includes specific procedures for handling hazardous pesticides.

Pesticide User's Guide. Bulletin XCM—85 (Colorado).

Safe Handling of Agricultural Pesticides. Leaflet 2768 (California).

Safe Handling of Pesticides. Publication MT8412—AG (Montana).

Soil Fumigation: How and Why It Works. Extension Publication 528 (Idaho).

Soil Fumigator's Manual. Publication EM 4387 (Washington).

Tank-Mixing Herbicides. Publication PNW 255 (Idaho, Oregon, Washington).

Insects

Cost Analysis of Alternative Insect Control Programs for Washington Potato Production. Publication XCO636 (Washington).

Costs of Suppressing Leaf Roll on Potatoes by Systemic Soil Insecticides. Publication XCO641 (Washington).

Economics of Cultural Methods to Suppress Green Peach Aphid as a Virus Vector. Publication XBO900 (Washington).

Economics of Various Soil Insecticide Treatment Programs for Controlling Green Peach Aphid. Publication XCO642 (Washington).

Idaho Potato Insect Handbook. Bulletin No. 505 (Idaho).

Management of Potato Insects in the Western States. Western Regional Extension Publication 64.

Pacific Northwest Insect Control Handbook. (Oregon).

Diseases and Disorders

Chemical Control of Plant Diseases. Publication 400W10 (New Mexico).

Compendium of Potato Diseases. American Phytopathological Society, St. Paul, MN, 1981, 125 pp.

Early Blight Prediction Model. In *Annual Report: Fungus and Bacterial Diseases Research,* 1982 (Colorado).

Pacific Northwest Plant Disease Control Handbook. (Oregon).

Plant Pathology. G.N. Agrios, Academic Press, New York and San Francisco, 2d ed., 1978, 703 pp.

Submitting Plants for Plant Disease Diagnosis. Publication 400W1 (New Mexico).

Nematodes

Collecting Samples for Nematode Analysis. Publication 110017 (Arizona).

Control of Nematodes. Publication 400W8 (New Mexico).

Cotton Rootknot Nematode. Publication 110423 (Arizona).

Differential Host Test for the Columbia and Northern Root–Knot Nematodes. Research Circular XC 0645 (Washington).

Plant Nematology, An Agricultural Training Aid. S. M. Ayoub, NemaAid Publications, P.O. Box 23058, Sacramento, CA 95823, revised 1980, 195 pp.

Sampling for Nematodes in Soil. Research Bulletin XB 0923 (Washington).

Weeds

Application of Herbicides through Irrigation Systems. Bulletin XOM–99 (Colorado).

Applied Weed Science. M. A. Ross and C. A. Lembi, Burgess Publishing Co., Minneapolis, 1985, 340 pp.

Chemical Weed Control. Publication 400A15 (New Mexico).

Chemical Weed Control for the Irrigated Areas of Arizona. Publication 183055 (Arizona).

Chemical Weed Control Guide. Extension Circular 301 (Utah).

Chemical Weed Control in Potatoes. Extension Bulletin 0842 (Washington).

Colorado Weed Control Handbook. Bulletin XCM–38 (Colorado).

Common Weed Seedlings of Montana. Publications 2L325, 2L326 (Montana).

Growers Weed Identification Handbook. Publication 4030 (California).

Herbicides Through Sprinklers. Current Information Series No. 369 (Idaho).

How to Identify Plants. H. D. Harrington and L. W. Durrell, Swallow Press, Chicago, 1957, 203pp.

Metribuzin for Potato Weed Control. Current Information Series No. 291 (Idaho).

Pacific Northwest Weed Control Handbook. (Oregon).

Responses of Potatoes and Weeds to Herbicides. Bulletin 844 (Washington).

Selective Chemical Weed Control. Leaflet 2827 (California).

Sprinkler Application of Herbicides. Publication EM 4532 (Washington).

Weeds of California. W. Robbins, M. Bellue, and W. Ball, State of California Documents and Publications, P.O. Box 1015, North Highlands, CA 95660, 1970, 547pp.

Weeds of Colorado. Bulletin 521A (Colorado).

Weeds of Wyoming. Bulletin 498R (Wyoming).

Wyoming Weed Control Guide. Bulletin 442R (Wyoming).

Testing Laboratories

Commercial Analytical Laboratories in California Available for Agricultural Testing. Leaflet 3024 (California).

Plant Disease Diagnosis Services. Publication 181138 (Arizona).

Publications of each state are available from that state's local extension offices or from the following addresses. Western Regional publications are available from these addresses in each state.

Arizona: Agricultural Communications, College of Agriculture, University of Arizona, Tucson, AZ 85721

California: Agriculture and Natural Resources Publications, University of California, 6701 San Pablo Avenue, Oakland, CA 94608-1239

Colorado: Bulletin Room, Colorado State University, Fort Collins, CO 80523

Idaho: Agricultural Communications, Ag Publications Building, University of Idaho, Moscow, ID 83843

Montana: Cooperative Extension Service, Montana State University, Bozeman, MT 59717

New Mexico: Bulletin Office, Dept. of Agricultural Information, Drawer 3AI, New Mexico State University, Las Cruces, NM 88003

Nevada: College of Agriculture Publications, University of Nevada-Reno, Reno, NV

Oregon: Bulletin Mailing Office, Industrial Building, Oregon State University, Corvallis, OR 97331

Utah: Bulletin Room, Utah State University, Logan, UT 84322-5015

Washington: Bulletin Dept., Cooperative Extension, Cooper Publications Building, Washington State University, Pullman, WA 99164-5912

Wyoming: Bulletin Room, P.O. Box 3313, University Station, University of Wyoming, Laramie, WY 82071

Sources of Chemicals for Bruise or Nematode Host Tests

J. T. Baker Chemical Co., 995 Zephyr Avenue, Hayward, CA 94544

Fisher Scientific Co., 2170 Martin Avenue, Santa Clara, CA 95050

Sigma Chemical Co., P.O. Box 14508, St. Louis, MO 63178

Glossary

abiotic disorder. a disease caused by factors other than a pathogen; physiological disorder.

annual. a plant that normally completes its life cycle of germination, growth, reproduction, and death in a single year.

apical dominance. growth of the bud at the apex of a stem or tuber while growth of all other buds on the stem or tuber is inhibited.

apothecia (plural). cup-shaped, spore-bearing structures produced by certain types of fungi such as *Sclerotinia*.

axil. the upper angle between a branch or leaf and the stem from which it grows.

axillary bud. a bud formed in an axil.

band application. the application of a material such as fertilizer or herbicide in strips, usually to the bed at planting or the side of the hill.

biennial. a plant that completes its life cycle in two years and usually does not flower until the second season.

biotic disease. a disease caused by a pathogen.

broadcast application. the application of a material such as fertilizer or herbicide to the entire surface of a field.

calcareous soil. soil containing high levels of calcium carbonate.

canker. a dead, discolored, often sunken area (lesion) on a root, stem, or stolon.

chlorophyll. the green pigment of plants that captures the energy from sunlight necessary for photosynthesis.

chlorosis. yellowing or bleaching of plant tissue that is normally green, usually caused by the loss of chlorophyll.

circulative virus. a virus that systemically infects its insect vector and usually is transmitted for the remainder of the vector's life; persistent virus.

control action guideline. a guideline used to determine whether pest control action is needed.

curing. holding potato tubers under warm, humid conditions that favor wound healing.

degree-day. a unit of measurement combining temperature and time, used in monitoring early blight (page 83).

determinate. having a growth pattern in which stems stop growing at a certain developmental stage.

developmental threshold. the lowest or highest temperature at which growth occurs in a given species.

dormancy. a state of inactivity during which the buds of potato tubers will not sprout.

drag-off. the practice of removing soil from the tops of potato hills before sprout emergence.

epidermis. the outermost layer of living cells on the surface of a plant or animal.

eye. a collection of several buds on the surface of a potato tuber, one of which will sprout and form a new stem when conditions are favorable (Figure 6).

field capacity. the moisture level in soil after saturation and runoff (Figure 14).

glycoalkaloid. a bitter-tasting compound present in potato foliage and in the epidermis of potato tubers.

host. a living organism that is invaded by a parasite and from which the parasite obtains its food.

indeterminate. having a growth pattern in which stems continue growing indefinitely.

infection. the entry of a pathogen into a host and establishment of the pathogen as a parasite of the host.

infestation. the presence of a large number of pest organisms in an area or field, on the surface of a host or anything that might contact a host, or in the soil.

inoculum. any part or stage of a pathogen, such as spores or virus particles, that can infect a host.

larva. the immature form of an insect, such as a caterpillar or maggot, that hatches from an egg, feeds, then enters a pupal stage.

latent. producing no visible symptoms (generally refers to an infection or a pathogen).

latent period. the time between when a vector acquires a pathogen and when the vector becomes able to transmit the pathogen to a new host.

leaf area index. the ratio between the total leaf surface area of a plant and the surface area of ground that is covered by the plant.

lenticels. natural openings in the surface of a tuber or stem, similar to leaf stomata, that can open and close and allow gas exchange (Figure 6).

lesion. a localized area of diseased tissue, such as a canker or leaf spot.

meristem. the tissue of plant growing points, the cells of which can divide continuously.

micropropagation. generation of new, disease-free potato plants from tiny pieces of meristem tissue.

microsclerotia. very small sclerotia, such as those produced by the Verticillium wilt fungus.

minituber. a small tuber produced under greenhouse conditions on a small potato plant generated by micropropagation.

mycelium (plural, **mycelia**). the vegetative body of a fungus, consisting of a mass of slender filaments called hyphae.

necrosis. death of tissue accompanied by dark brown discoloration, usually occurring in a well-defined part of a plant, such as the portion of a leaf between leaf veins or the xylem or phloem in a stem or tuber.

node. the slightly enlarged part of a stem where buds are formed and where leaves, branches, and eyes originate (Figure 4).

nonpersistent virus. a virus that is carried on the mouthparts of its insect vector and is lost after the vector feeds once or a few times; styletborne virus.

nymph. the immature stage of insects such as aphids and psyllids that gradually acquires adult form through a series of molts without passing through a pupal stage.

parasite. an organism that lives in or on the body of another organism, the host, from which it derives its food without killing the host directly; also used in this book to describe an insect that spends its immature stages in the body of a host, which is killed just before the parasite pupates.

pathogen. a disease-causing organism.

perennial. a plant that can live three or more years and flower at least twice.

periderm. several layers of corky cells located on the outside of the epidermis of a potato tuber and containing high amounts of suberin.

persistent virus. a virus that systemically infects its insect vector and is usually transmitted for the remainder of the vector's life; circulative virus.

petiole. the stalk connecting a leaf or leaflets to a stem (Figure 4).

pheromone. a substance secreted by an organism to affect the behavior or development of others of the same species.

phloem. the food-conducting tissue of a plant vascular system (Figure 5).

physiological disorder. a disorder caused by factors other than a pathogen; abiotic disorder.

phytotoxicity. the ability of a material such as a pesticide or fertilizer to cause injury to plants.

postemergence herbicide. in this book, an herbicide applied after potatoes emerge; also, an herbicide applied after target weeds emerge.

preemergence herbicide. in this book, an herbicide applied before potatoes emerge; also, an herbicide applied before target weeds emerge.

primary bloom. the first production of flowers on a potato plant, occurring after 8 to 12 leaves have been formed on the mainstem and generally coinciding with the beginning of the tuber growth phase.

pupa. the nonfeeding, inactive stage of an insect during which the larva is transformed into an adult.

reservoir. the site where a pest population or quantity of inoculum can survive in the absence of a host crop, and from which a new crop may be invaded.

resistant. able to withstand conditions harmful to other strains of the same species.

rogue. to remove diseased plants from a field.

rolling. mechanical crushing of potato vines to hasten vine death, sometimes used synonymously with vine-killing.

rosetting. abnormal growth caused by certain pathogens in which new potato foliage is stunted and tightly bunched.

rugose. a rough appearance of leaves in which veins are sunken and interveinal tissue raised, caused by certain virus infections.

russeting. thickening of the periderm on tubers of russet cultivars that occurs after vine senescence.

sclerotium (plural, **sclerotia**). a compact mass of hardened mycelium that serves as a dormant stage for some fungi.

secondary bloom. a second production of flowers on a potato plant, occurring at the end of the mainstem of an indeterminate cultivar; secondary bloom may occur on a determinate cultivar at leaf axils along the mainstem.

secondary infection. infection by a pathogen that enters the host through an injury caused by another pathogen.

secondary outbreak. the increase of a nontarget pest to harmful levels following a pesticide application, caused by destruction of natural enemies that normally control the nontarget pest.

secondary spread. the spread of a pathogen within a field after the initial or primary infection.

secondary stems. stems formed by stolons that emerge from the soil.

seed leaf. leaf formed within a seed and present on a seedling at germination; cotyledon.

seed piece. portion of a potato tuber containing at least one eye that is planted to produce a new potato plant.

specific gravity. the ratio of the density of a substance to the density of pure water; specific gravity of potato tubers is used as a measure of their dry matter content.

spraing (sprain). reddish brown spots, rings, or arcs in tuber tissue caused by tobacco rattle virus; corky ringspot.

sprout. the new stem formed from the eye of a potato tuber.

sprout inhibitor. a chemical applied to potato vines or to stored tubers to prevent sprouting.

stolon. the underground stem of a plant, the end of which may form a tuber.

stoma (plural, **stomata**). natural opening in a leaf surface that serves for gas exchange and water evaporation and has the ability to open and close in response to environmental conditions.

stroma. a compact, usually spore-producing structure formed from fungal mycelium on the surface of a host.

styletborne virus. a virus that is carried on the mouthparts of its insect vector and is lost after the vector feeds once or a few times; nonpersistent virus.

suberin. a waxy substance, resistant to microbial attack, formed in the corky cells of periderm layers.

suberization. the formation of periderm layers on the cut surfaces or wounds of potato tubers.

systemic. capable of moving throughout a plant or other organism, usually in the vascular system.

tensiometer. a device, used for measuring soil moisture, that consists of a closed, buried tube that develops a partial vacuum as the soil dries out.

tertiary bloom. the third production of flowers that occurs at the end of the growing stem of an indeterminate potato cultivar.

tolerance level. maximum percentage of a disease or pest symptom allowed during field inspections for certification of a seed lot; levels are different with each field generation and may vary from state to state.

tolerant. used to describe a cultivar that is able to grow and produce an acceptable yield when infected by a pathogen.

toxin. a poisonous substance produced by a living organism.

translocated herbicide. the preferred term for a systemic herbicide.

transpiration. the evaporation of water from plant tissue, mostly through stomata.

treatment threshold. the level of a pest population at which a control measure is needed to prevent eventual economic injury to the crop.

true leaf. any leaf produced after the seed leaves (cotyledons).

tuber. an enlarged, fleshy, underground stem with buds capable of producing new plants.

tuberization. the formation of tubers at the ends of stolons; tuber initiation.

vascular ring. a thin area of potato tuber tissue between the cortex and the medulla in which vascular tissue is concentrated (Figure 6).

vascular system. the system of plant tissues that carries water, mineral nutrients, and products of photosynthesis throughout the plant, consisting of xylem and phloem (Figure 5).

vector. an organism capable of transmitting a pathogen to a host.

vegetative growth. growth of leaves, roots, and stems, including tubers, as opposed to flowers or fruits.

vein banding. dark brown discoloration of the veins on the undersides of potato leaflets caused by potato virus Y.

wound healing. suberization.

xylem. tissue that conducts water and mineral nutrients from the roots to the rest of the plant (Figure 5).

Index of Names for Diseases and Disorders